Naturalists' Handboo

Animals of sandy shores

PETER J. HAYWARD

With illustrations and plates by the author

Pelagic Publishing
www.pelagicpublishing.com

Published by **Pelagic Publishing**
www.pelagicpublishing.com
PO Box 725, Exeter, EX1 9QU

Animals of sandy shores
Naturalists' Handbooks 21

Series editors
S. A. Corbet and R. H. L. Disney

ISBN 978-1-78427-039-1

Digital reprint edition of ISBN 0-85546-293-0 (1994)

British Library Cataloguing in Publication Data
A catalogue record for this book is available from the British Library.

Contents

Editors' preface

The assemblage of animals living in sandy shores is richer than it might at first appear, and it offers wonderful opportunities for ecological exploration without the need for expensive equipment. This book introduces the natural history of the community and provides keys that will enable readers to name the animals they find. It describes practical approaches for behavioural and ecological studies, including the survey and monitoring of populations. Local investigations of this kind form an essential basis for planning the conservation of sandy shore habitats, which are important both in their own right and as feeding grounds for birds.

S.A.C.
R.H.L.D

Acknowledgements

I am grateful to Jan Greengo for her skill with the keyboard, and to Nathalie Yonow for critically reviewing the first draft of this book.

P. J. H.
July 1992

1 Introduction

Shorelines exert an attraction for most people. Seaside holidays are a tradition in all temperate and tropical countries with coastlines, and the beach is the most enduring popular image of the seashore. For the holidaymaker a sandy beach is ideal for seaside recreation, free from rocks, seaweed and unfamiliar animals. Yet, sandy shores support a surprisingly rich fauna, largely unseen by the casual visitor, and naturalists, when able to study beaches undisturbed by the pressures which the most attractive and most accessible of them suffer, find many interesting fields for research.

Sandy shores are found on most parts of Britain's coastline. Long stretches of soft sand backed by sparsely vegetated dunes are a feature of the East Anglian coast, parts of South Wales, North Devon and Somerset, Lancashire, Northumberland, western Scotland and the Hebrides. South Devon, Cornwall, West Wales and many parts of Scotland are graced by tiny secluded coves where sandy beaches are hidden by flanking cliffs or headlands. Along the southern coast of England much of the accessible beach consists of narrow strands of coarse sand, often grading into shingle, pebbles or large boulders. The shore may be loosely defined as that part of the land which is covered by the sea at high tide and exposed at low tide. The nature of a shore, whether a soft, golden beach, or a narrow, shingly strand, a weedy rock platform, or a storm beaten headland, is determined by the interaction of the sea with the geology and topography of the adjoining land.

Coastlines are not static; they change seasonally and annually in response to varying physical factors, and are profoundly influenced by long-term geological cycles. Continental land masses consist of light, low density rocks supported by underlying plates of heavy, dense rocks. These two major rock types exist in a state of dynamic equilibrium, referred to as isostasy (fig. 1). Periodically, volcanic events or glaciations disturb this balance; massive outflows of high density lava, or the growth of ice caps perhaps several km thick, increase the weight of the land mass, which sinks lower into its underlying support. Similarly, rapid melting and retreat of a glacier lightens a land mass, which consequently moves upward. Land masses are constantly adjusting to such events, and this isostatic adjustment (fig. 1) results in continually fluctuating sea levels. Further, as ice caps advance and retreat, more or less water is bound up as ice, and sea levels rise and fall accordingly, a phenomenon referred to as eustasy, or eustatic adjustment. Britain's coastline has undergone considerable change over the last 100, 000 years, as the glaciers advanced and retreated, and is

isostasy: the dynamic equilibrium between the light rocks of the continental land masses, and the underlying heavy rocks of the earth's mantle

eustasy: the dynamic equilibrium between land and sea level

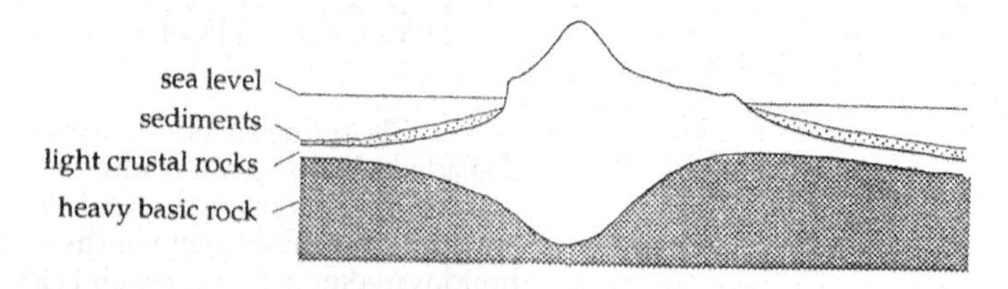

Fig. 1. Isostasy: the equilibrium between light rocks of the continental crust and heavy basic rocks of the underlying mantle.

still responding to both isostatic and eustatic phenomena. Successive rises and falls of sea level are marked by raised beaches, rock platforms, and riverine terraces, which may be particularly conspicuous on some coasts. On parts of the North Kent, Essex and Suffolk coasts sea level is still rising, and considerable areas of land have been lost to the sea in historical times. Elsewhere, such as around the Solway Estuary, the sea has retreated as a result of isostatic rising of the land.

A second major factor determining the character of a coastline is its geology. Soft rocks, and unconsolidated sediments such as sand and gravel, are swiftly eroded by the sea, giving rise to abrupt shorelines, such as the chalk cliffs of south-east England, or the sandy cliffs of Norfolk and parts of Hampshire. Harder rocks erode more slowly, to give low, rounded headlands and beaches of smooth cobbles; the granite shores of western Scotland and western Ireland are good examples. Coastlines are difficult to classify. Broadly speaking, rising, or emergent, coastlines may be classified as "depositing coastlines" (fig. 2), where sand and mud deposited by the sea exceed that eroded and carried away. However, not all depositing coastlines are necessarily emergent. Sinking, or submergent, coasts are a major category of "eroding coastlines", where projecting headlands and cliffs are continuously eroded by the sea but where drowned valleys and the deepening sea floor accumulate the eroded material (fig. 2). Coastlines, their form and evolution, are considered in detail by Steers (1969a)*, who has also produced excellent accounts of the history and structure of Britain's coasts (Steers, 1962, 1969b). Sandy beaches are necessarily shores of deposition, and the rest of this chapter will deal specifically with these.

Coastal erosion and deposition both result from the impact of waves on the shore. Wind blowing across a water surface causes water molecules to move in a circular fashion,

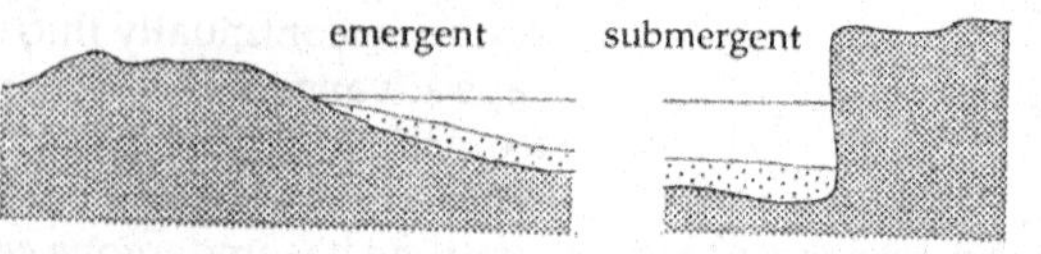

Fig. 2. Sediments (stippled) accumulated on two types of depositing coastline.

* References cited under the author's name in the text appear in full in the list of references on p. 98.

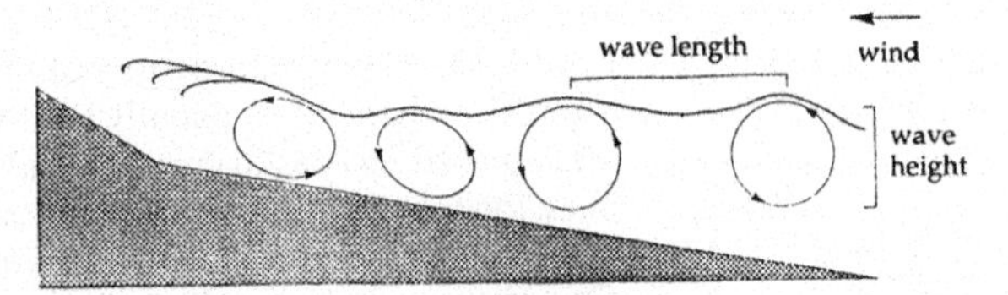

Fig. 3. The movement of water particles in breaking waves.

and the passage of a wind is thus marked by alternating peaks and troughs, or waves. Wind strength governs the speed of movement of the water molecules, and wave length, height and velocity are all closely interdependent (fig. 3). Waves slowly lose energy, becoming longer and flatter as they travel from their point of origin; powerful waves produced by oceanic storms eventually reach distant shore lines as a "swell". The velocity, height and persistence of a swell depend on the "fetch" – the distance of open sea over which a swell has passed. Southwestern shores of Britain and Ireland experience southwesterly swells which have travelled over thousands of miles of ocean surface, while eastern England has swells with a comparatively minor fetch. Waves generated by winds or storms close to the coastline are referred to as "seas". Both swells and seas are modified on approaching land (fig. 3). As the sea shallows, friction with the bottom distorts the orbital water movement, wave crests develop, and when wave height exceeds water depth, the wave breaks. On a gently shallowing shore this will occur long after the waves' energy has been dissipated; on a steep, abruptly shallowing coast waves are equally abruptly slowed, and break with destructive force. The erosive power of such waves is enhanced by sand, pebbles, and even large cobbles or boulders, swept up from the bottom and thrown onshore by the breaking wave. Waves may be differentially slowed, or bent (refracted) by uneven bottom topography, or by projecting headlands (fig. 4). Energy is lost rapidly at these points and more slowly over gently shelving, adjacent bays so that wave fronts approach the shore obliquely. At such places the returning waves move along the shore, as long shore drift, which at points may develop into a fast offshore undertow, or rip current.

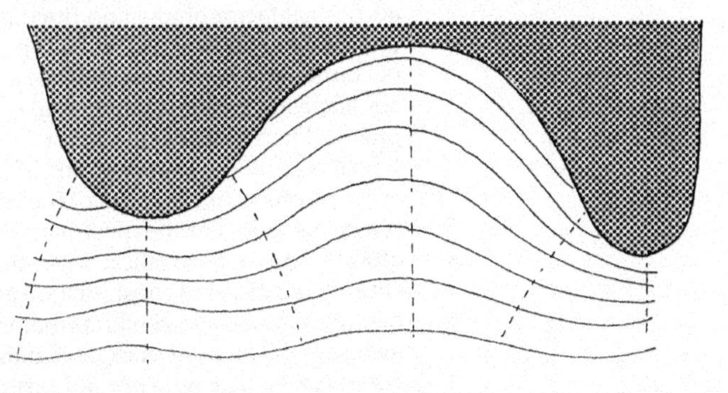

Fig. 4. Refraction of waves approaching a coast. Dotted lines indicate paths of wave crests.

Depositing shores are built from material eroded from elsewhere. Most usually this is from coastal sources, transported by coastal currents and long shore drift, but it may include material dumped by rivers as they drain into the sea. The deposits vary widely, from coarse pebbles to fine, silty muds, depending on wave velocity and its rate of decrease as it approaches the shore. As wave energy decreases, heaviest materials are deposited first; gravel and shingle is followed by sand, silt, and finally the smallest clay particles. Depositing shores are sometimes referred to by biologists as "mobile shores", a particularly appropriate term as deposits are continually sorted and graded by wave action to give a wide variety of potential habitats. Exposed and steeply dipping shores, subject to plunging waves, tend to develop narrow pebble or shingle strands from which all finer material has been washed out and carried away. Waves break abruptly, piling up deposits on the beach and accentuating its steepness. On low gradient shores, where wave energy dissipates more slowly, coarse material is dropped offshore and finer particles are carried inshore to develop a sandy beach. Sudden changes in shore topography resulting from storms, landslips or artificial events such as the building of breakwaters or offshore dredging, or more gradual changes due to local current patterns, may all affect the nature of beaches. Natural spits or bars may appear or disappear quite abruptly, or gradually develop through time, or persist unchanged for long periods. Depositing shores are inherently unstable, with particular consequences for the faunas inhabiting them.

This book is concerned primarily with sandy shores. High energy, plunging shores of gravel and shingle, formerly thought to be devoid of life, may in fact have a very specialized fauna, but this is generally not very rich. It is difficult to study, and is still the province of the specialist. Intertidal mud flats have a particular fauna, typically composed of very few, but very abundant species. Mudflat habitats are most common in estuaries, where the additional factor of low and fluctuating salinity poses particular ecological problems. These two types of shore are, of course, simply the two ends of a very wide spectrum of aquatic habitats, but their special characteristics make it practical to consider them apart from the sandy shores which occupy the broadest part of the spectrum.

Beach sand consists of the resistant products of coastal erosion. For the most part, these are chipped or ground fragments of silica. Some beaches consist almost entirely of silica, but most include an admixture of other materials, especially shell fragments, and as shelter increases the amount of organic material, in particular, may also increase. The type of sand forming a particular beach

Φ : the Greek letter phi

Table 1. *The Wentworth Scale for the classification of sediments*

Type of sediment	Diameter, mm	Φ units
Boulder	>256	> –8.0
Cobble	64 - 256	–6.0 - –8.0
Pebble	4 - 64	–2.0 - –6.0
Granule	2 - 4	–1.0 - –2.0
Very coarse sand	1 - 2	0 - –1.0
Coarse sand	0.5 - 1	1.0 - 0
Medium sand	0.25 - 0.5	2.0 - 1.0
Fine sand	0.125 - 0.25	3.0 - 2.0
Very fine sand	0.0625 - 0.125	4.0 - 3.0
Coarse silt	0.0312 - 0.0625	5.0 - 4.0
Medium silt	0.0156 - 0.0312	6.0 - 5.0
Fine silt	0.0078 - 0.0156	7.0 - 6.0
Very fine silt	0.0039 - 0.0078	8.0 - 7.0
Coarse clay	0.00195 - 0.0039	9.0 - 8.0
Medium clay	0.00098 - 0.00195	10.0 - 9.0

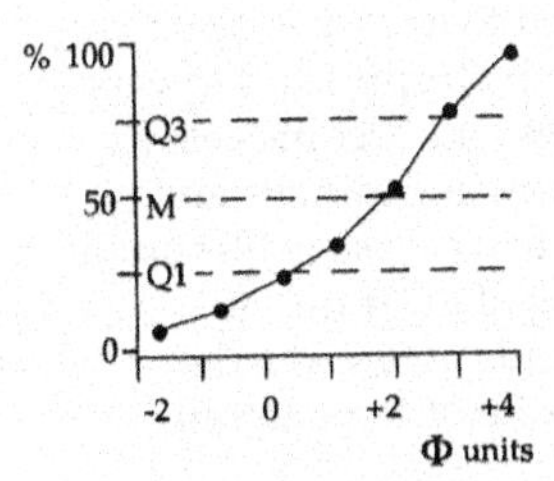

Fig. 5. Cumulative frequency curve for a sandgrain particle analysis.
M =median Φ value.
Q1 = first quartile.
Q3 = third quartile.
Quartile deviation value = (Q3-Q1)/2

allows an immediate insight into its ecology. Sand particle size correlates with physical environmental factors affecting the beach, and with the distribution of the beach fauna. Grain size analysis is an important first step in studying a beach, and in making ecological comparisons between beaches. Sand is defined as material with a grain diameter ranging from 0.063 mm to 2 mm. Larger material is classified successively as gravels, pebbles and cobbles, and finer material is graded into various categories of silt and clay. These categories are formalised in a scale, the Wentworth Scale (table 1), and sand ecologists use a standard series of sieves, with mesh sizes coinciding with the units of the Wentworth Scale, in establishing the grain size distribution for a particular sand sample. The percentage weight of a sample retained by each sieve is plotted as a cumulative curve (fig. 5). Most often results are expressed as phi (Φ) units, a simplified notation in which the Wentworth Scale measures are transformed to $-\log_2$, to give a series of single values (table 1). For comparative purposes a sand sample is usually categorised by its median Φ number – the value at which 50% of the sample is coarser and 50% finer (fig. 5). Another useful value is the Φ quartile deviation – the difference between the Φ values for 25% and 75% of the sample. This is a good measure of the range of grain sizes present in the sample.

The cumulative curve derived from grain size analysis yields important information about the sample (fig. 6). A steep curve, with a small Φ quartile range, indicates a well sorted sand in which all grains are very similar in size. Such a sand has high porosity. In an arrangement of evenly packed spheres pore space is at a maximum at 26%; thus, a uniformly graded sand has a high pore space and a high potential water content. Well sorted sands are characteristic

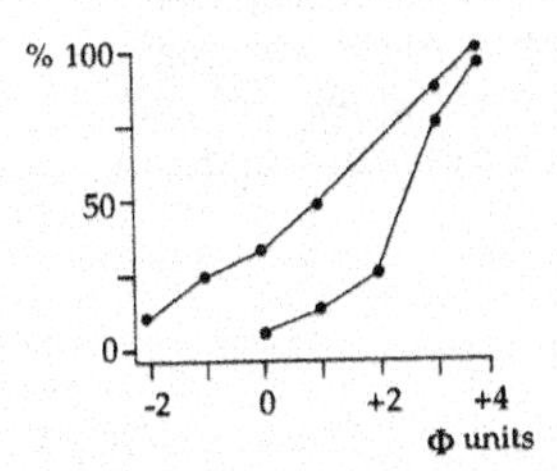

Fig. 6. Cumulative frequency curves for a mixed sand (left) compared to a well sorted sand (right).

of exposed shores. A shallow cumulative curve, and a correspondingly broad Φ quartile deviation (fig. 6), indicate a poorly sorted sand, composed of particles of many different sizes. Such sands are associated with low profile, sheltered beaches.

Porosity is not the most important aspect of water/substratum balance for intertidal animals. In a well sorted, coarse sand water drains swiftly away as the tide falls; its permeability is thus said to be high. In poorly sorted sands porosity is low, but the water retaining ability of the sand is enhanced by capillary action, and permeability is thus low. However, in the finest sands, porosity may be at a maximum through uniformity of grain size, but capillary action is increased by the small size of the particles; permeability is again low, and the deposit will have a high water retaining capacity. In both very fine deposits, and mixed, poorly sorted sands in which interstitial water content is high, a phenomenon referred to as thixotropy may be recognised. Pressure applied to a thixotropic sand lowers its viscosity. In effect, the sand liquefies and becomes easy to penetrate, a fact of importance to burrowing animals and to shore-feeding birds. Pressure applied to a coarse, well sorted sand causes it to harden as the packing of its grains is tightened, driving out water. These sands, referred to as dilatant sands, become increasingly difficult to penetrate with pressure.

thixotropy: the tendency for mixed, porous sands to liquefy when pressure is applied

dilatancy: the tendency for well sorted sand to harden when pressure is applied

No sandy beach is uniform in composition. The profile may show a varying gradient between high and low tide marks, and exposure to swell, waves and long shore drift will vary along the beach. Sheltered corners with fine, silty sand may grade towards coarser, more evenly sorted sand in the middle of the bay, and the composition of the sand fauna will change accordingly. A sample of sand, sieved and analysed, gives an immediate and quite precise insight into the ecology of the habitat at the sampling point. In practice, and especially for comparative studies, it is usual to produce grain size analyses and cumulative curves for successive stations along a transect line, at standard intervals related to tidal level. A series of transects spaced along a beach will give an accurate picture of the varying ecological characteristics of the beach. Techniques for beach surveying and establishing transects are described in detail by Jones (1980).

2 Living in sand

Intertidal sand is a difficult habitat in which to live. It is fundamentally unstable, and subject to regular small-scale disturbance, and to irregular but often frequent large-scale disturbance resulting from storms. It also suffers from often rapid fluctuations in physical and chemical environmental characteristics. Subtidal sandy deposits are equally unstable but are not subjected to the additional stresses resulting from twice-daily inundation and exposure by the tides.

The most obvious disadvantages to sandy shore living are a lack of refuge from predators, and exposure to atmospheric conditions during low tide. An incoming tide brings with it fish, crabs and other predators. Some, such as flatfishes, cannot survive total emersion, and must move in and out with the tide. Others, like the Necklace shells, *Lunatia* species (pl. 4.5, 4.6), and most crabs, survive low water periods by burrowing into the sand, emerging once more to hunt over its surface as the tide flows in. At low tide all aquatic animals, predators and prey alike, become potential food for a host of terrestrial animals, most noticeably gulls and wading birds, but also insects, and even a few mammals. Consequently, most sandy shore animals, except predatory species, live below the surface during all states of the tide, and at low tide every species seeks refuge in temporary or permanent burrows and a sandy beach can appear devoid of life.

For the most part, the physical and chemical characteristics of sandy shore habitats are relatively stable during periods of immersion. The overlying body of sea water is locally uniform, so that temperature, salinity, oxygen and food levels are constant over the whole beach and there is little of the microclimatic variation found on rocky shores where, for example, pools of water at different tidal levels drain and are refilled at different rates, and thus may have different values and ranges for these factors. However, as the tide recedes from a sandy shore, the habitat is subject to atmospheric effects, whose influence may result in sharp and sometimes dramatic variations in environmental factors. Temperature is among the most important of these. Seawater temperatures vary little between day and night, and seasonal variations tend to be slow and gradual, allowing animals to respond through behavioural changes. However, at low tide air temperature becomes critically important to intertidal animals, and on sandy beaches the habitat, from the surface to a depth of several centimetres, can experience large variations in temperature during a single tidal cycle. The effect of air temperature depends on the time of day at which low tide occurs, and this differs from one part of the British coast to another. The most profound effects result from low tides at around midday and midnight; in the former case, summer low tides may be particularly damaging, exposing animals to peak noonday temperatures, while winter midnight low tides expose them to the hazards of frost. Low temperatures tend to be more damaging than high. Seawater freezes at -1.8°C, a value only rarely reached in coastal waters around Britain. However, winter air temperatures frequently drop as low as this, and during prolonged cold spells freezing temperatures and the formation of surface ice may lead to the death of many sedentary organisms such as Razor shells (pl. 4.1), Cockles (pl. 8.6) and Tellins (pl. 4.7). Motile animals escape the dangers of winter cold simply by migrating down shore to beyond the reach of the tides; thus, crabs, shrimps, brittle stars and Sand Gobies are absent from the intertidal stretches of sandy shores for most of the winter. Deep burrowing species such as the Lugworm, *Arenicola* (pl. 7.2), and the Gaper Clam, *Mya* (pl. 8.7, 8.8), are probably safe from all but the most profound freeze–ups.

The effect of unusually cold weather is a simple physical one, in which body fluids freeze, causing cell and tissue damage, and eventually rupture of cell walls, and even of the body wall. The effects of high temperature are not so direct, but rather are related to the influences of other physical factors, most particularly oxygen. Only surface-living animals exposed at the lowest tides are likely to succumb to heat stress, and even in their case it is not clear that temperature *per se* is the immediate cause of death. For most intertidal sand-dwellers high temperatures may be deleterious through their effect on oxygen levels. At the sand surface, summer sunshine stimulates vigorous photosynthetic activity by unicellular algae, to the extent that small surface pools may become saturated with oxygen. However, high temperatures also promote rapid bacterial growth, and on beaches with a high organic content – decaying seaweed, for example – the demand for oxygen is so great that, just a few mm below the surface, the oxygen content of the interstitial water is swiftly depleted and the habitat becomes practically anaerobic. The Lugworm *Arenicola* (pl. 7.2), and other large, burrowing polychaetes such as *Owenia fusiformis* (pl. 3.3), have a blood haemoglobin with a particularly high affinity for oxygen, and thus have a reserve which enables them to overcome short-term deficiencies in oxygen supply. *Arenicola* ventilates its burrow with fresh sea water during high tide, and at low tide, as oxygen decreases sharply, it may draw bubbles of air into its burrow and respire by absorbing oxygen through its body wall, provided it remains wet. *Arenicola* can survive for up to nine days under completely anaerobic conditions and there is some evidence (Dales, 1958) that during this time it switches to anaerobic metabolism, deriving energy from the breakdown of glycogen without oxygen. The tube-building worm *Owenia* has been found to survive anaerobiosis for up to 21 days (Dales, 1958), but glycolysis does not occur in this species, which simply survives oxygen depletion by lowering its energy demands and remaining completely quiescent. Generally, oxygen depletion is a problem especially important to the smallest animals, which are critically affected by interstitial water quality. Larger burrowing animals may cope with the short-term problems of the low tide period by adjusting their position in the sand, or modifying their behaviour.

‰ : parts per thousand

Salinity is a physical environmental factor of importance to all marine organisms. The salinity of coastal waters around the British Isles ranges from 31.0‰ to 35.2‰, with a seasonal variation of 1–2‰, dependent on the amount of evaporation from surface waters during the summer, and the amount of dilution by winter rains and river runoff. Generally, the fauna of a shore reflects local salinity regimes. Many intertidal animals can regulate the salt levels of their body fluids and are said to be euryhaline, tolerating a wide range of salinities; others are stenohaline, adjusted to only a narrow range of salinities. Marine stenohaline animals are usually osmoconformers, whose body fluids have the same salt content as ambient seawater. However, many intertidal animals, such as the ubiquitous Green Shore Crab, *Carcinus* (pl. 5.3), are osmoregulators, able to control their body's salt and water balance. Salinity is most likely to become a problem for the sand fauna at low tide. During high tide interstitial water in the top few cm of sand has the same salinity as the water above, and the period between successive low tides may be sufficient to allow circulation of fresh, full strength seawater to a considerable depth in the sand. At low tide, however, rain and freshwater runoff from the upper shore may drastically reduce salinity in the top levels of the sand. Bivalve molluscs counter this threat simply by closing their shells. Large burrowing animals may retain water of suitable salinity in their burrows, or resist osmotic flooding by covering the body with a layer of mucus. However, many small animals living in the top levels of the beach must simply adapt to regular salinity fluctuation. On some shores, such as the well-studied beach at Kames Bay, Millport, freshwater seeps from the upper shore through the sand, so that some areas of the beach have interstitial salinities permanently lower than that of the overlying seawater. The polychaete worms *Arenicola* (pl. 7.2) and *Nephtys*, which are both osmoconformers and unable to tolerate freshwater runoff, are excluded from such areas and replaced instead by the Ragworm *Nereis diversicolor* (fig. 7), an osmoregulating euryhaline species. Plotting the distribution of the larger

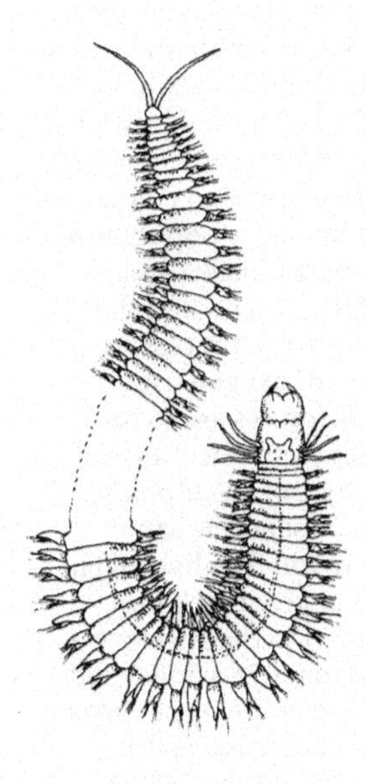

Fig. 7. *Nereis diversicolor.*

polychaetes over a beach, in terms of numbers recorded down a series of transect lines, is thus a good way of estimating interstitial salinity regimes.

We saw in Chapter 1 that sand particle size distribution is a sensitive indicator of the ecological conditions pertaining on a particular beach. Very fine, poorly sorted sands have a high water retaining capacity, usually enhanced by their high content of organic detrital particles. Permeability is low, and such beaches often have pools of standing water throughout the entire low tide period. Oxygen depletion may be a severe problem at all states of the tide on the very finest grained beaches. As a general rule, if the percentage of particles of less than 0.25 mm diameter exceeds 10% of a sand, then the oxygen concentration of its interstitial water will be less than 20% of the air saturation level, and will drop swiftly during low tide periods. Fine sands provide the greatest grain surface area for the attachment of micro-organisms – such as bacteria and diatoms – and their rich organic content is capable of supporting large populations. On Nova Scotia beaches, Dale (1974) found that sediment with a mean grain diameter of 0.01 mm contained 10% organic carbon, and 5000×10^6 bacteria per gm dry weight, while sand with a mean grain diameter of 0.2 mm contained only 0.1% organic carbon, and 200×10^6 bacteria per gm dry weight. Fine sands usually support the richest fauna, although oxygen depletion, and variations in salinity and pH, result in steep gradients of faunal change through the upper layers of the beach. A sharp boundary occurs at 5 to 15 cm depth, where the colour of the sand changes abruptly from yellow to black; this boundary is usually marked by a thin grey layer (fig. 8). Above the grey layer the interstitial water contains sufficient oxygen for the contained fauna, and to oxidize all of the organic waste products of the immense populations of micro-organisms. Below the boundary the black sand is essentially anoxic; free oxygen is totally absent, and the microfauna must survive through anaerobic processes. Bacteria still thrive, but in the absence of oxygen use fermentation, or other chemosynthetic processes, to break down organic compounds. Many bacteria reduce sulphate, nitrate or carbonate ions to produce hydrogen sulphide, ammonia or methane, which give black sand the same unpleasant smell as sticky estuarine muds. The hydrogen sulphide reacts with iron in the sand to give black iron sulphides; as these are gradually carried to the surface by burrowing animals they are oxidised to ferric oxide, which imparts the yellow colour characteristic of the upper layers. There is a difference in electrical potential between the reduced ions of black sand and the oxidised ions of the upper, yellow layers, and this is used by ecologists to express the oxygen demand, or redox potential, of the black

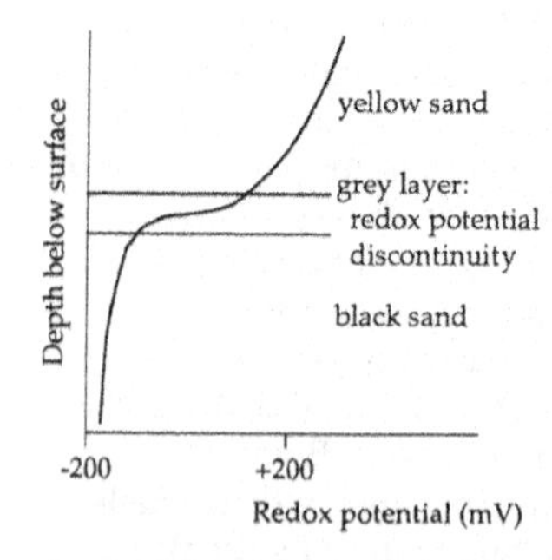

Fig. 8. Sand profile showing the change of redox potential with depth.

redox potential: a measure of the oxygen debt of black sand (or mud) environments

sand (fig. 8). In the total absence of free oxygen the redox potential (Eh) drops to -200 millivolts, compared with a value of +400 mV at the surface. The grey boundary layer marks the transition from reducing to oxidising conditions and is termed the redox potential discontinuity, or RPD. In coarse, well drained sand there is no black layer as all organic material is rapidly oxidised. As sand becomes progressively finer, and less well sorted, with increasing shelter, so the black layer moves closer to the surface. At the furthest end of the spectrum black muds are entirely anoxic and all free-living organisms in them must depend on anaerobic metabolism. The depth of the black sand boundary is thus a good indicator of ecological conditions prevailing on a particular beach, although it should be noted that this is not fixed. The depth of the boundary is greatest during the winter months but decreases during the summer as bacterial populations bloom and oxygen demand rises.

An excellent introduction to sandy shore ecology is given by Brafield (1978), and more detailed accounts may be found in Eltringham (1971) and Brown & McLachlan (1990).

3 Sandy shore animals

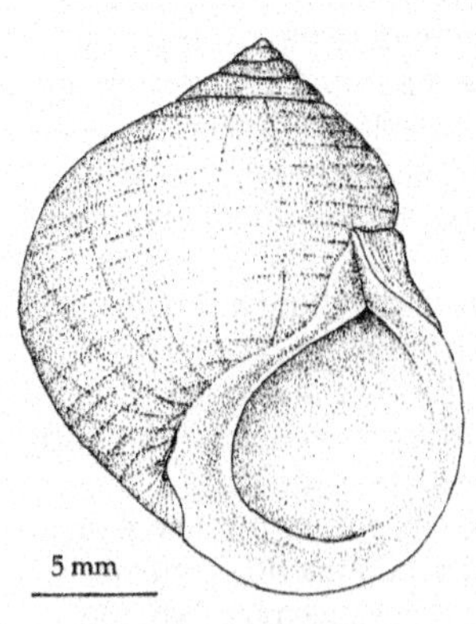

Fig. 9. *Littorina littorea* (L.).

Many different kinds of animal can be found on a good, sheltered sandy shore. Some species seem almost ubiquitous: the Green Shore Crab, *Carcinus maenas* (pl. 5.3), the Common Periwinkle, *Littorina littorea* (fig. 9), and the Common Starfish, *Asterias rubens* (pl. 6.1), can all be found on sand, as well as on rocky shores, on mud, and in estuaries. However, such species are few and, in general, the fauna of sandy beaches is quite different from that of rock and mud. As a rule, each animal group is represented by few species, but some of these will be amazingly abundant. The bivalve *Angulus tenuis* (pl. 4.7), for example, may achieve densities of 3000 per square metre. Comparing numbers of species of different types of animal between sandy and rocky shores, though over-simplistic, neatly demonstrates the distinction between the two faunas (table 2). Rocky shores provide habitats for a wealth of encrusting and attached organisms, such as sponges, hydroids, bryozoans and barnacles, which support a rich diversity of grazing and predatory animals, particularly snails, crustaceans and echinoderms. The uniformity of the sandy shore, in contrast, results in a lower overall species count; certain groups are excluded altogether, but bristle worms and bivalves are overwhelmingly more diverse in sand than on rock.

The intertidal fauna of sandy beaches is strictly marine in origin. The strandline and the narrow supralittoral fringe above it are populated by a mixture of animals with both aquatic and terrestrial origins. Strandline debris harbours large populations of sandhoppers and seaslaters whose evolutionary pedigrees are essentially marine, and

Table 2. *Comparison of the numbers of species of different animal groups on muddy, sandy and rocky shores (from Eltringham, 1971)*

	Wareham Point Kingsbridge S. Devon	Mill Bay Kingsbridge S. Devon	Lannancombe Bay S. Devon
Taxonomic group		Numbers of species	
	Mud	Sand	Rock
Porifera	0	0	3
Cnidaria	2	2	5
Turbellaria	1	0	0
Nemertea	1	1	0
Annelida	9	7	2
Sipunculida	1	0	0
Crustacea	4	6	14
Mollusca	15	6	15
Echinodermata	0	1	1

equally large numbers of beetles, flies, bugs, mites and spiders, all of which originate from terrestrial habitats. A few of these latter are specially adapted to the strandline habitat, while others utilise it during only part of their life cycle. The strandline constitutes a separate habitat, in many ways even harsher than intertidal sand and distinctly different from it. Its environmental characteristics, and heterogeneous fauna, demand separate treatment and it will not be considered further here.

epifauna: animals living on the surface of the substratum

infauna: animals living within the substratum

Two major categories of sand-dwelling animals may be recognized: the epifauna and the infauna. The term epifauna refers to animals which live on the surface of the sand. Most are non-resident and must move in and out with the tide; a few survive low tide periods by burying themselves in the sand, becoming temporary members of the infauna, and emerging to feed when the tide returns. The epifauna includes flatfish, sand eels, weaver fish, the Sand Star *Astropecten* (pl. 6.4) and the brittle star *Amphiura brachiata* (pl. 6.3), and predatory molluscs such as *Lunatia* (pl. 4.4, 4.5) and the shelled sea slug *Acteon* (pl. 4.2). However, although many epifaunal animals may be specially adapted for life on sand, very few can tolerate an intertidal existence. The boundary between epifauna and infauna becomes rather blurred when one considers some of the small crustaceans, such as shrimps, opossum shrimps (or mysids), cumaceans and amphipods. Most of these appear to be non-resident epifauna, moving in and out with the tide, but many join the infauna as the tide retreats and some, particularly among the amphipods, have restricted, and quite clearly defined, intertidal distributions.

a cumacean

macrofauna: "large animals": species which do not pass through a 1 mm sieve mesh

meiofauna: "smaller animals": species which pass through a 1 mm sieve mesh, but are retained by a 0.05 mm mesh

microfauna: "small animals": animals less than 0.05 mm in diameter

The infauna is the larger of the two categories of sandy shore animals and includes many specially adapted intertidal species. It is conveniently divided into three major types – the macrofauna, the meiofauna and the microfauna – defined simply by reference to the animals' sizes. Macrofaunal animals are large enough to be retained by a 1 mm sieve; the meiofauna pass through a 1 mm sieve but are retained by a mesh of 50 μm (0.05 mm), while the microfauna are organisms of less than 50 μm. There is a good ecological basis for this classification. While macrofaunal animals must dig permanent or temporary burrows in the sand, or constantly tunnel through it, the meiofauna lives in the spaces between sand grains, does not disturb them as it passes between, and does not create or maintain burrows. The microfauna comprises the smallest protozoans, which may live between sand grains, or on their surfaces, or in films of water surrounding them. Study of the microfauna requires special microbiological techniques. Although it is perhaps of profound importance in conditioning the chemistry of the interstitial habitat, it is not otherwise the concern of the field biologist.

a mysid

The meiofauna is the most varied of the three groups of infaunal animals, and although its total weight, or biomass, nowhere approaches that of the macrofauna, it often far exceeds it in terms of numbers of individuals. Almost every kind of marine invertebrate animal can be found among the meiofauna; some are specially adapted species of groups more usually found among the macrofauna, such as molluscs, while others belong to groups which are predominantly meiofaunal in habit, such as the harpacticoid copepods. In summer, meiofaunal populations may be boosted by temporary residents, such as the newly settled spat of cockles and tellins which live an essentially interstitial life for their first few weeks. On the Dutch coast up to 70,000 cockle spat have been recorded in a single square metre of intertidal sand. The resident meiofauna (pl. 1) includes foraminiferans, ciliate protozoans, hydroids, flatworms, rotifers, gastrotrichs, nematodes, archiannelids, polychaetes, molluscs, tardigrades and numerous types of crustaceans. Nematodes have sometimes been suggested as the most diverse and abundant animal group and on most sandy beaches usually comprise the bulk of the meiofauna. Interstitial crustaceans include a few isopods, and many ostracods, but the overwhelming majority are harpacticoid copepods, whose slender bodies are particularly suited to an interstitial existence, and which are among the most important of the meiofaunal groups. Moore (1979*a*) conducted a survey of meiofaunal communities on six Isle of Man beaches and recorded animal densities as high as 7655 individuals per 10 cm^2 of sand surface; nematodes were the dominant group at each sampling station, with harpacticoid copepods a close second. The curious tardigrades, or water bears, were also common at some stations, especially species of *Batillipes* (pl. 1.1). Regrettably, most *Batillipes* species are less than 0.5 mm long, and their startling appearance is only revealed by a good compound microscope.

a nematode

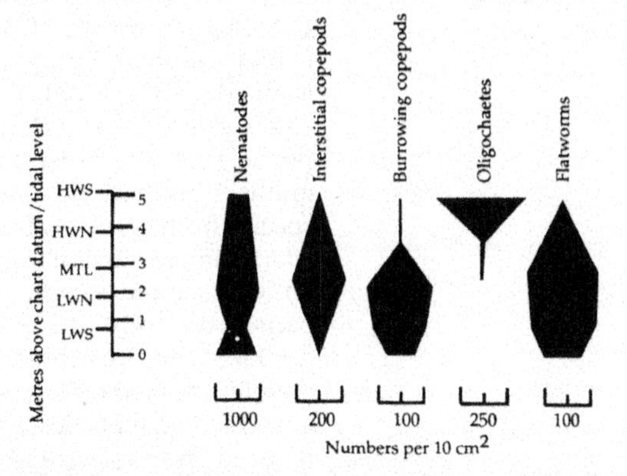

Fig. 10. Down shore distribution of some groups of meiofauna (after Moore 1979*a*).

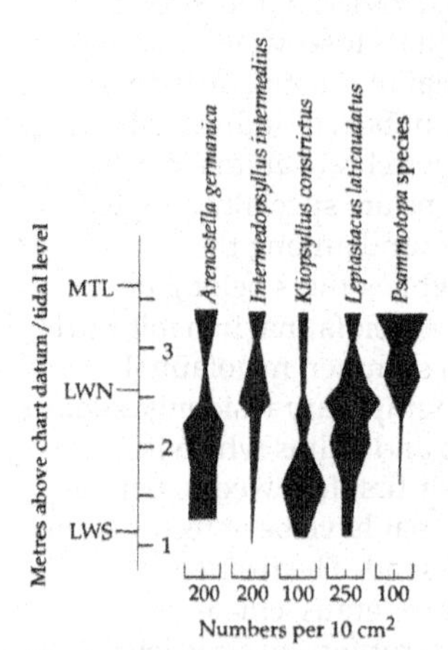

Fig. 11. Low shore zonation of some meiofaunal harpacticoids (after Harris 1972).

Meiofaunal communities flourish in both subtidal and intertidal sands and muds. Between the tides, the most important factor governing their distribution is the level of water saturation in the sand. Medium to fine sand, with a minimum 80% saturation during low tide, will support the richest meiofauna. As saturation drops, in coarser sands, a critical point is reached, at about 50% saturation, below which the density of the meiofauna decreases sharply, and relative proportions of different types of animal begin to change. This may be seen on a single beach, for example that at Port Erin studied by Moore (1979*a*), where down shore transects through regions of different exposure, and hence different grain size distributions, show differing proportions of the main meiofaunal animals (fig. 10). There are even clear zonation patterns within each group, as Harris (1972) and Moore (1979b) demonstrated with harpacticoid copepods (fig. 11). These two examples illustrate both vertical (down shore) and horizontal (along shore) zonation in the meiofauna. While pore space, and hence water saturation, is clearly a major factor influencing these distribution patterns, others, such as temperature and salinity, will also have important effects. The majority of the meiofauna lives within 15 cm of the sand surface, but may occur shallower or deeper according to surface temperatures and the amount of freshwater runoff at low tide. There is also evidence that the overwintering portion of the fauna tends to move deeper into the sand during the winter months.

The many different groups of animals contributing to the meiofauna have certain characteristics in common. Nematode worms and harpacticoid copepods, with their slender body forms, seem preadapted to life in the interstices between sand grains, and most other groups represented have similar, elongate or decidedly worm-like shapes. The Acochlidiacea is an order of sea slugs, the most familiar of which are *Microhedyle* (pl. 1.6) and *Hedylopsis* (pl. 1.7), both with a maximum size of 1–3 mm. The prosobranch sea snail *Caecum* (fig. 12), with its curious cylindrical shell, 1 mm long, is scarcely recognized as a snail. Most interstitial ciliates tend to be larger than other ciliate protozoans, and may bear plates or spines which perhaps serve to resist the crushing effect of moving sand grains. Wright (1983) describes and illustrates more than 50 species from South Wales beaches. Among other groups, adaptation to a meiofaunal life has involved a reduction in body size, often accompanied by loss or simplification of parts of the body. The tiny coelenterate *Halammohydra* (pl. 1.4), which resembles the freshwater *Hydra* but is only remotely related to it, has a slender body with only seven, long, feeding tentacles, and in certain archiannelid worms reduced body size has resulted in fewer cells and the loss of

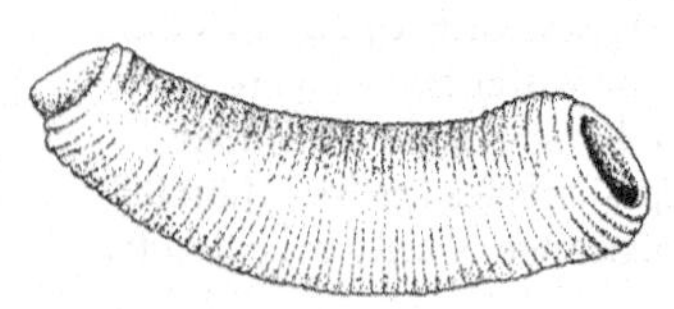

Fig. 12. *Caecum imperforatum.*

half of the usual complement of testes and excretory nephridia. In a majority of meiofauna the body surface is covered with cilia, and the animals move between sand grains by ciliary swimming or gliding. Most species produce only a small number of eggs, which are often sticky and so not easily washed away and lost, or which develop in specialized cocoons, often brooded by the adult. As a general rule, meiofaunal animals do not have free-swimming planktonic larvae, and reproductive modifications such as copulation, the transmission of sperm in packets (or spermatophores), or even hermaphroditism, all serve to prevent wastage of eggs and sperm.

Some meiofaunal animals feed on organic detritus, others on bacteria, while a majority feed on the diatom flora of the beach. Despite their small sizes they are thus important in sandy shore food chains, and as primary consumers of the smallest food particles are essential to the ecology of the shore. Young plaice and dab feed voraciously on meiofaunal harpacticoids, and also cockle and tellin spat, during their first weeks of life, and for brown shrimps and sand gobies the meiofauna is the major food source during much of their life.

Study of the meiofauna demands careful collection and preparation of samples, and a good microscope is essential. Identification is often difficult, although useful field surveys can be carried out on the distribution and abundance of major animal groups, and grading by size within each group will separate many species. The most important modern handbook to the meiofauna is that of Higgins & Thiel (1988), while the classic review of Swedmark (1964) is still a good starting point. McIntyre (1969) provides a useful introduction to meiofaunal ecology, and Hicks & Coull (1983) review the ecology of meiofaunal harpacticoids. Techniques for meiofaunal studies are given in detail by Holme & McIntyre (1984).

Macroinfaunal animals are by definition burrowers. Their burrows may be only temporary structures, or they may be permanent, with the animal occupying the same position in the sand throughout its adult life. Bioturbation is a term which describes the continual reworking – overturning and mixing – of sands and muds by infaunal animals, and indeed the endless burrowing, and the passage of sediment through the guts of generations of animals, are important ecological processes in sand and mud habitats. The beach macrofauna includes specialized species of echinoderm such as the Sea Potato, *Echinocardium cordatum* (pl. 6.2), the brittle star *Amphiura brachiata* (pl. 6.3), and the sea cucumber *Leptosynapta inhaerens* (pl. 7.4). There are burrowing prawns, such as *Upogebia* and *Callianassa*, with limbs modified for digging, and the Masked Crab, *Corystes cassivelaunus* (pl. 5.1), which uses its elongated antennae to maintain a respiratory passage to the surface. A host of smaller crustaceans, particularly amphipods, occupy tiny, impermanent burrows in the top few centimetres of sand. However, the predominant animal types in this habitat are the polychaetes, or bristle worms, and the bivalve molluscs.

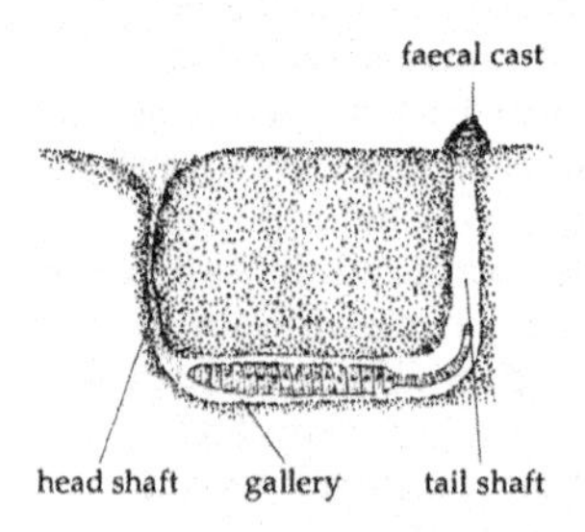

Fig. 13. Vertical section showing *Arenicola* in its burrow.

The polychaetes include a number of fast-moving predators, tunnelers rather than burrowers, and others which build only temporary burrows, but also numerous species which maintain substantial permanent burrows. The most familiar example, the Lugworm, *Arenicola marina*, builds a U-shaped burrow which it enlarges and deepens as it grows (fig. 13). The worm spends most of its time in its living gallery, at the bend of the U, which it ventilates by drawing a current of water in through the permanently open tail shaft. Periodically, it moves backwards up the tail shaft, voiding a faecal spiral of fine sand and waste materials. *Arenicola* feeds by ingesting sand from the base of the head shaft; some of this is excavated by the worm in the process of enlarging its burrow, but some is supplied by a continual rain of fine sand and detritus down the shaft, which appears on the surface as a partially blocked, saucer-shaped depression. *Sabella pavonina*, the Peacock Worm (pl. 3.1), uses fine particles of silt bound together with mucus to form a thin, tough tube with which it lines its burrow. The Mason Worm, *Lanice conchilega* (pl. 3.2), does essentially the same, but uses large sand grains and shell fragments to make a remarkably resistant tube, which extends for 30 cm or more downwards into the sand. Generally, the more sedentary the worm, the more permanent its burrow. The tube-building species are usually filter feeders, deriving their food directly from the seawater rather than from the sand itself. An exception is *Pectinaria koreni* (pl. 3.4), which builds a stout, tusk-like shell of cemented sand grains, but actually lives upside down, feeding on organic material in the sand.

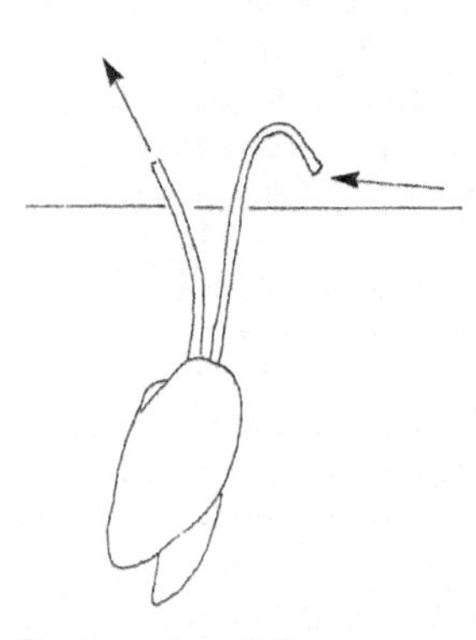

Fig. 14. A deposit-feeding bivalve with long, separately extensile siphons.

Mode of life is equally varied among bivalves, and there is good correlation between life style, shell shape and the structure of the twin siphons which bivalves use to maintain water circulation through their mantle cavities (fig. 14). The Wedge Shell, *Donax vittatus* (pl. 4.6), is a fast-moving, tunneling species with a slender, pointed shell, a narrow profile and a long, pointed foot. *Donax* lives close to the surface on moderately exposed beaches and is frequently washed out of the sand by surf. At the water's edge on a rising tide, Wedge Shells washed free by the surf will bury themselves again within seconds, and juveniles of 1 cm length or less can be seen to jump, so rapidly is the foot extended. Tellins (*Angulus* species) are also shallow burrowers, with thin, pointed shells which allow very swift burrowing. Most species of *Angulus* tend to lie on one side in the sand, and the two valves of the shell, unusually in

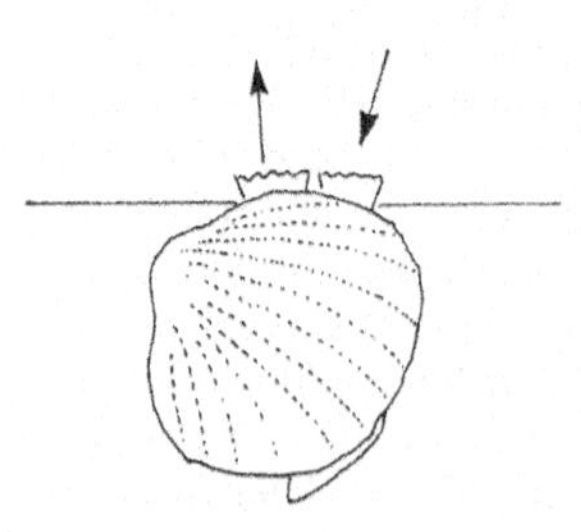

Fig. 15. The Cockle, a shallow burrowing suspension feeder, with short, partly fused siphons.

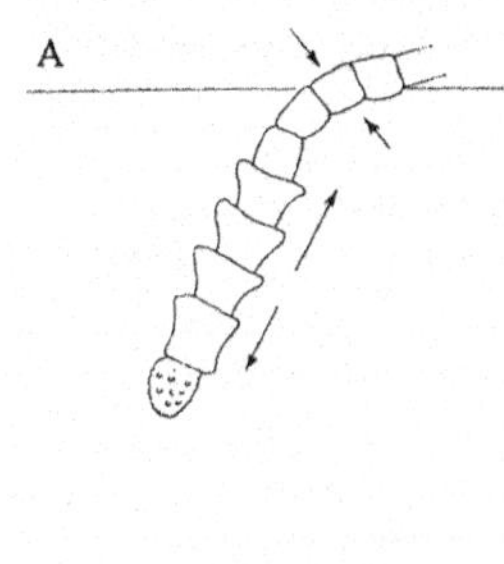

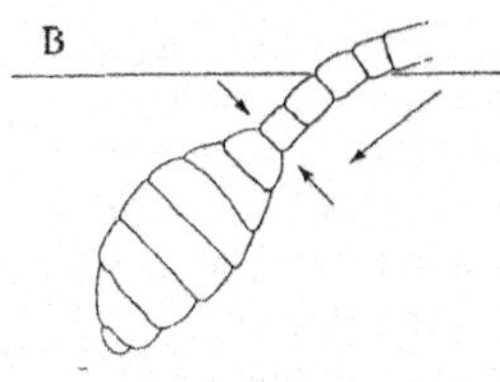

Fig. 16. The burrowing cycle in *Arenicola*.
A. Elongation of head end and formation of penetration anchor
B. Formation of terminal anchor as the worm pulls itself into the sand.

Fig. 17. The digging cycle in a burrowing bivalve.
A. Foot probes, siphons open.
B. With siphons closed, valves close expelling water, and tip of foot forms terminal anchor.
C. Anterior retractor muscle contracts, top side of shell digs in.
D. Posterior retractors contract, bottom side digs in.
E. At rest.

sand-dwelling bivalves, tend to be slightly dissimilar. The stream-lined shapes of the Razor Shells *Ensis* (pl. 4.1) and *Solen* (fig. VIII.4) are adapted for rapid, deep burrowing, and these animals occupy permanent, vertical burrows from which they are rarely dislodged. The squat, bulbous shapes of cockles are typical of shallow burrowing species, whose shape helps them maintain position with their posterior, siphon-bearing ends level with the sand surface (fig. 15). The heavy-shelled Gaper Clam *Mya arenaria* (pl. 8.7) lives deeply buried in permanent burrows; unlike Razor Shells it does not move in its burrow, but instead has greatly elongated, fleshy, and only partly retractile, siphons which reach to the surface of the sand.

Burrowing techniques are basically the same in both polychaetes and bivalves. Penetration of sand during burrowing may require considerable pressure, difficult to sustain in animals lacking a skeleton. In polychaetes and bivalves this lack is made good by muscles acting against internal body fluid pressures to provide an efficient hydrostatic skeleton. In polychaetes the hydrostatic skeleton comprises the fluid in the main body cavity, and the circular and longitudinal muscles of the segmented body. In bivalves the body cavity is reduced to a small chamber surrounding the heart, but the muscular flexible foot contains large blood spaces, collectively termed the haemocoel, and constitutes the animal's hydrostatic skeleton. Functional convergence between these two animal types is paralleled in their digging and burrowing cycle, which is fundamentally the same in both cases. Once a worm, such as *Arenicola*, has penetrated the sand, contraction of circular muscles in the middle region of its body increases internal hydrostatic pressure, and a "penetration anchor" is formed as segments behind the head swell outwards (fig. 16). The penetration anchor holds the animal in place as the head end elongates, probing deeper into the sand. Contraction of muscles just behind the head then causes it to swell, forming a bulbous "terminal anchor" (fig. 16). This gives the worm sufficient purchase to enable it to draw its body forwards into the burrow. This burrowing cycle is basically the same in bivalves, but is a slightly more complex process involving precise opening and closing of shell valves, one result of which is a softening of the substratum by water forcefully expelled by the animal. In *Angulus tenuis*, studied in detail by Trueman and others (1966), the middle region of the foot forms a penetration anchor as its tip probes into the sand (fig. 17); the siphons close as the foot reaches full extension, and as its tip swells to form a terminal anchor the shell valves are swiftly closed by rapid contraction of their adductor muscles, expelling water from between the valves into the sand. Paired muscles in the foot then contract to draw the animal downwards; these are not quite in phase, the anterior muscles acting ahead of the posterior, resulting in a seesaw movement which very effectively digs the shell deep into the sand.

Opportunities for feeding specialisation are comparatively few in intertidal sandy habitats. Among polychaetes and bivalves one may broadly divide feeding into two types – suspension feeding and deposit feeding. Suspension feeders derive their nutriment from planktonic organisms and detritus carried in the overlying water, while deposit feeders derive theirs directly from the sand. *Sabella* and *Lanice,* for example, partly emerge from their tubes during high tide periods and filter food from the water with their tentacle crowns, while *Arenicola* digests organic material, bacteria and diatoms from the sand it swallows. Deposit feeding bivalves generally have flexible, extensible siphons which vacuum food particles from the sand surface, while suspension feeders have short, fused siphons specialized simply to maintain continuous inhalant and exhalant water currents (figs. 14, 15). In reality, the distinction between these two types of feeding can become rather blurred, in that some suspension feeders may use their tentacles actually to select edible particles from the sand surface, and, besides, are often filtering out particles from deposits which are simply resuspended with every tidal cycle. Similarly, deposit feeding bivalves may often function also as filter feeders. Echinoderms, however, seem to be largely deposit feeders on sandy beaches, using their tube feet to pick up food. On the whole, macroinfaunal carnivores are few. The Catworm, *Glycera alba* (II.47, p. 56), is an active infaunal predator, feeding on other polychaetes and on small crustaceans, but the equally common *Nereis* and *Nephtys* are perhaps as much scavengers as predators.

4 Distribution and zonation

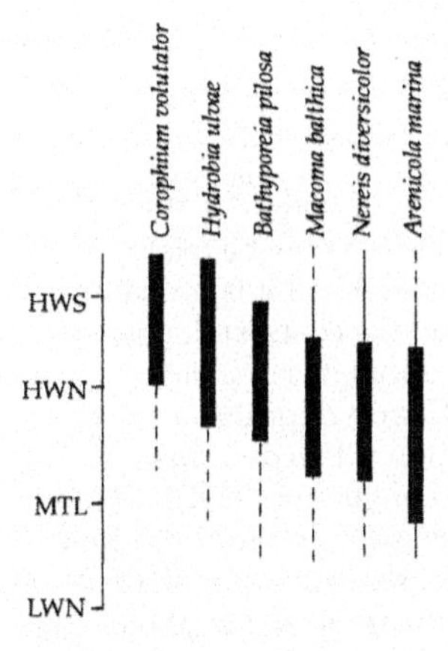

Fig. 18. Downshore zonation of some common sandy shore animals.

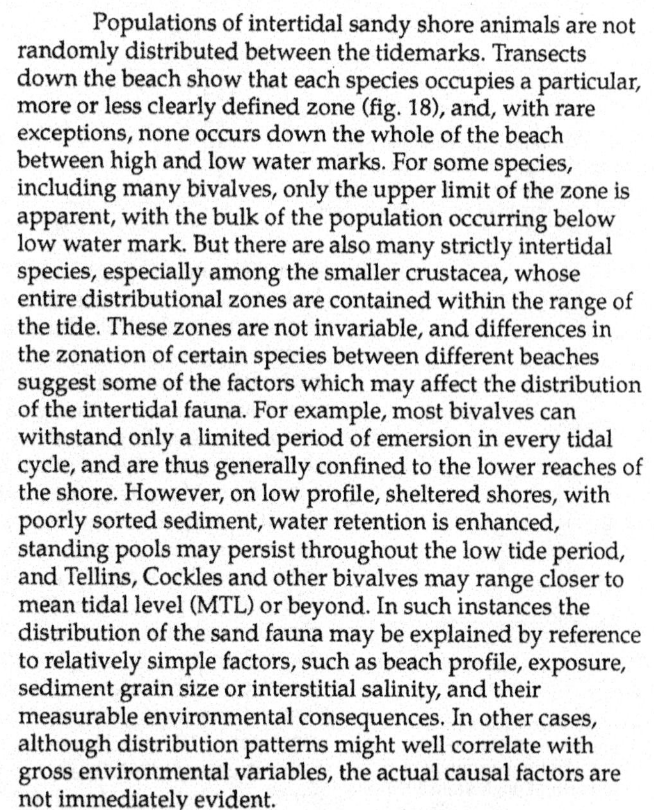

Populations of intertidal sandy shore animals are not randomly distributed between the tidemarks. Transects down the beach show that each species occupies a particular, more or less clearly defined zone (fig. 18), and, with rare exceptions, none occurs down the whole of the beach between high and low water marks. For some species, including many bivalves, only the upper limit of the zone is apparent, with the bulk of the population occurring below low water mark. But there are also many strictly intertidal species, especially among the smaller crustacea, whose entire distributional zones are contained within the range of the tide. These zones are not invariable, and differences in the zonation of certain species between different beaches suggest some of the factors which may affect the distribution of the intertidal fauna. For example, most bivalves can withstand only a limited period of emersion in every tidal cycle, and are thus generally confined to the lower reaches of the shore. However, on low profile, sheltered shores, with poorly sorted sediment, water retention is enhanced, standing pools may persist throughout the low tide period, and Tellins, Cockles and other bivalves may range closer to mean tidal level (MTL) or beyond. In such instances the distribution of the sand fauna may be explained by reference to relatively simple factors, such as beach profile, exposure, sediment grain size or interstitial salinity, and their measurable environmental consequences. In other cases, although distribution patterns might well correlate with gross environmental variables, the actual causal factors are not immediately evident.

The Lugworm is probably the commonest macroinvertebrate inhabitant of intertidal soft sediments around the whole of the north-west European coastline. It is a significant part of the intertidal marine food web, being an important food for flatfish, shrimps, gobies, and carnivorous worms, as well as for hosts of wading birds. *Arenicola* can be found in immense numbers on suitable beaches, reaching densities as high as 70 m^{-2}. On the vast sand flats of the Dutch Wadden See, Lugworms comprise more than 20% of the total macroinvertebrate biomass (Beukema and de Vlas, 1979), with a mean ash-free (organic) dry weight of 5 gm m^{-2}, higher than that of all other polychaete species together. The bulk of the *Arenicola* population on any shore is found in a dense band below MTL, except in late summer when a second band appears on the upper shore (fig. 19). Population densities correlate with sediment grain size, being highest in fine muddy sands, declining as the sediment coarsens, and the worms are usually completely absent from the finest muds. Most muds are found in

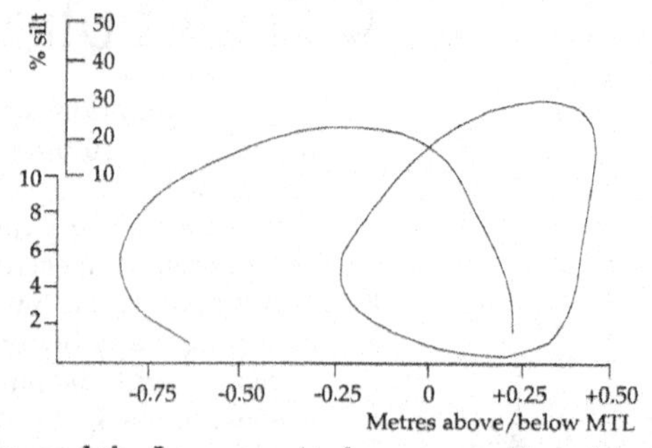

Fig. 19. Summer distribution of juvenile (right) and adult (left) *Arenicola marina* in relation to tidal level and silt content of sand (after Beukema and De Vlas 1979).

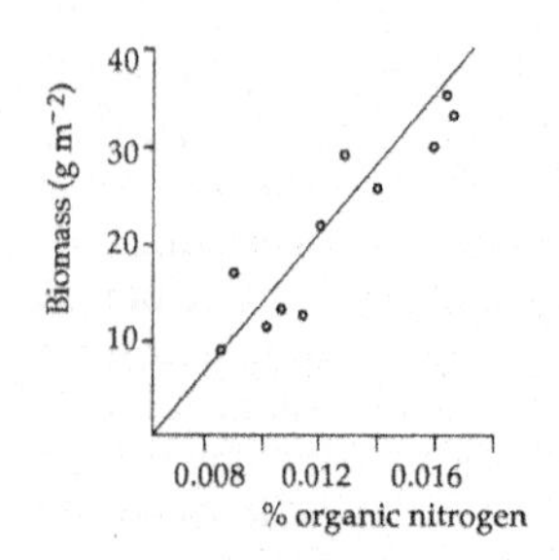

Fig. 20. *Arenicola marina* biomass in relation to the organic nitrogen content of sand (after Longbottom 1970).

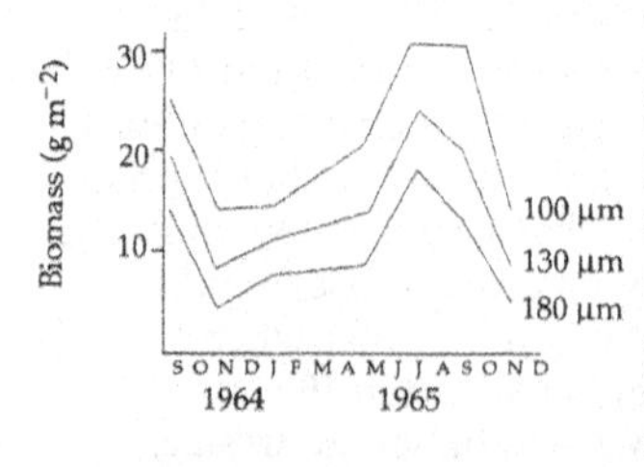

Fig. 21. Yearly variation in *A. marina* biomass, in three grades of sand (after Longbottom 1970).

estuaries, and the Lugworm's absence may be related to its lowest salinity tolerance (about 24‰), but it is probably excluded from both the finest and the coarsest deposits simply because in such loose material it is unable to maintain a permanent burrow. Coarse deposits are also unsuitable because, although a deposit feeder, *Arenicola* is unable to ingest large particles. Longbottom (1970) plotted the distribution of *Arenicola* at ten sites on the North Kent coast, from Margate to the Swale, recording population densities in terms of numbers of worms, and total biomass, per m^2. *Arenicola* biomass was found to correlate with median particle size of the sediment. In the finest deposits, with median particle size less than 80 µm, *Arenicola* was absent, except for juvenile worms living in the top few cm. As median particle size exceeded 200 µm – "fine sand" on the Wentworth Scale – *Arenicola* biomass declined sharply. Populations were greatest in sands with a median particle diameter of 100 µm. However, in grades of sand between the two extremes, 100-200 µm, Longbottom (1970) found a direct correlation between *Arenicola* biomass and organic content of the sand (fig. 20). The latter also correlates with median particle size, and the importance of organic content could be neatly demonstrated by plotting the seasonal variations in Lugworm biomass for three grades of sand (fig. 21).

The summer high shore *Arenicola* zone results from the upshore movement of juveniles. These leave the burrows of the females, where they hatched and underwent early larval development, at just 0.3 mm long, when they are just recognisable as juvenile polychaetes by their three bristle-bearing segments (fig. 22). The migration of these larval worms is probably passive: they are carried by the tide to upshore nursery areas of very fine sediment, and simply deposited along with other fine material. The young *Arenicola* overwinter in these nurseries, not actually burrowing but living in the top few cm of sediment, where they probably provide a rich source of food for small wading birds. In early spring the young lugworms, now 6–15 mm long (fig. 23), move down shore again and resettle in the adult habitat, where they construct their first permanent burrows. This second migration is also aided by the tide, although the young worms actively swim, leaving the sediment at night enveloped in transparent gelatinous tubes, and can make successive attempts at migration until the tide carries them to their preferred area (Farke and Berghuis, 1979). The question of how post-larval lugworms recognise their preferred habitat, and thus maintain the zoned distribution of the adult population, remains unanswered, and seems ripe for an experimental investigation.

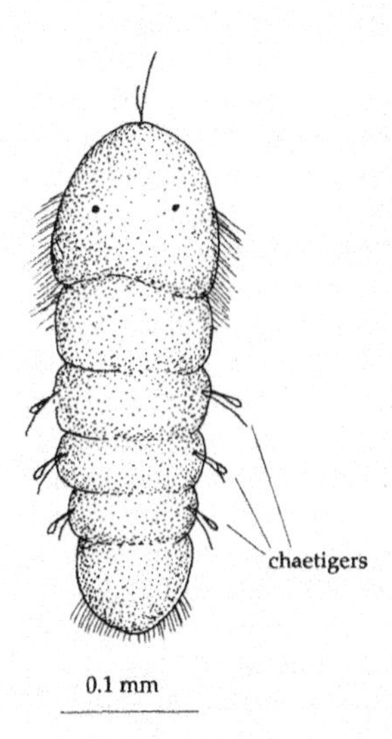

Fig. 22. *Arenicola marina:* 3-chaetiger stage (after Newell 1948).

Down shore zonation is displayed even more strikingly by small crustaceans, which are abundant, if not immediately evident, on all but the most exposed sandy shores. Among the smallest are the meiofaunal harpacticoid copepods studied by Moore (1979a, b) and Harris (1972). Their populations maintained zoned distributions on the beach throughout the year (fig. 11, p.16), some even occupying specific depth zones within the sand. Meiofaunal animals live permanently within the sand, feed and reproduce there, and display no daily or seasonal migratory behaviour. Eggs and larvae are generally brooded, or attached to sand grains, and hatch as advanced juveniles which do not undergo a planktonic larval infancy. Consequently there is every reason to suppose that meiofaunal species should have no difficulty in maintaining station in their optimum intertidal habitat. Most macroinfaunal invertebrates are less bound to the substratum. Many, especially among the bivalves, have planktonic larval stages, and it has been suggested (Warwick, 1989) that one good evolutionary reason for this is the high risk of predation on eggs and embryos by the meiofauna. A planktonic stage also allows the growing larva to capitalise on energy-rich planktonic foods, while avoiding the very real risk of falling prey to feeding adults. While numerous small crustaceans feed on detritus or sand-encrusting micro-organisms within the sand, others, like the isopod *Eurydice* (pl. 2.1), are hunting carnivores which seek their food on the sand surface or in the overlying water column. Further, while many small infaunal crustaceans brood their eggs and embryos, rather than trust them to the plankton, most must leave the sand and swim in order to find a mate. Swimming entails great risk of predation and thus usually occurs during the hours of darkness. These nightly swimming crustaceans are popularly referred to as the "surf plankton". They were first studied in detail by Elmhirst (1931) and Watkin (1939, 1941), in Kames Bay, Millport, from night-time plankton hauls made across the bay at approximately weekly intervals through a whole year. In describing the "night tidal migrant crustacea" Watkin (1941) distinguished between resident species which emerged at night from the sand, and others which moved

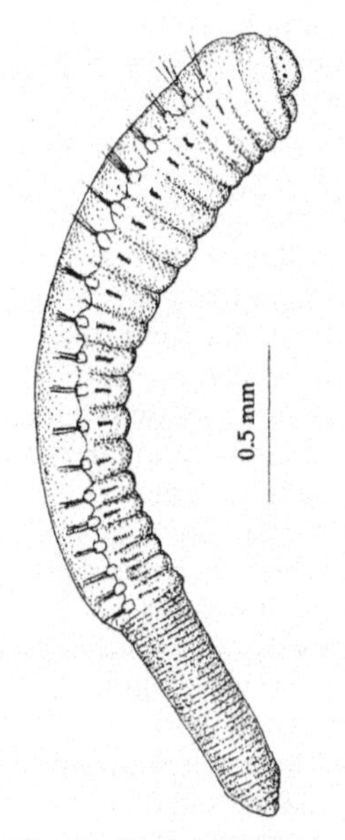

Fig. 23. *Arenicola marina*, Benham larva (after Newell 1948).

inshore, or were carried in by the tide, from adjacent subtidal habitats. The residents, especially species of *Bathyporeia* (pl. 2.3), *Pontocrates* (pl. 2.6) and *Urothoe* (pl. 2.2), were caught in greatest numbers over areas of the bay where they were known to be most abundant in the sand, and thus seemed to maintain their zoned distribution during swimming excursions. While resident populations included juveniles, males and females of all age classes, the incoming migrants, such as *Gammarus locusta* (fig.VI.34) and *Atylus swammerdami* (fig.VI.41), were represented almost entirely by juvenile specimens, and were probably involuntary migrants.

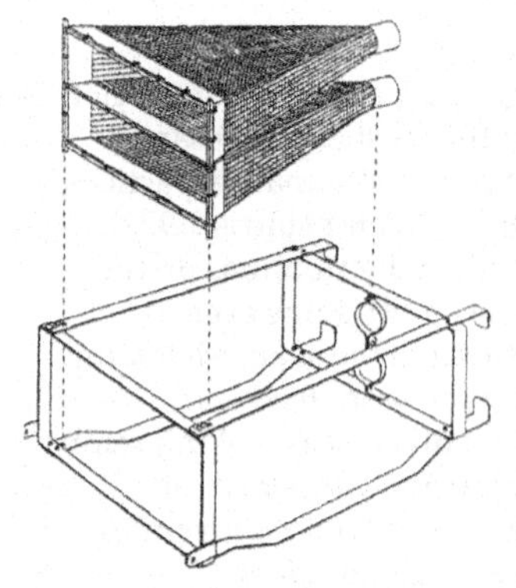

Fig. 24. Two-stage surf plankton net (after Coleman and Segrove 1955).

Coleman and Segrove (1955) devised a plankton net for quantitative sampling of night swimming crustacea (fig. 24). It had a rigid frame, incorporating runners which allowed it to be drawn across a sandy bay in water of about 0.5 m depth with minimum disturbance of sediment. A unique feature was its pair of nets, which sampled the upper and lower parts of the water column independently. Their results agreed with those of Elmhirst (1931) and Watkin (1939, 1941), while showing that whereas some species were distributed throughout the water column, others were most abundant either close to the bottom or in the surface layers. Fincham (1970) employed the Coleman-Segrove net in a survey of the night-swimming amphipods on Port Erin beach, Isle of Man. He sampled four times each month over a 19-month period, at the two neap and two spring tides of each lunar month. The samples comprised a single tow across the beach at a fixed hour – 2300 hrs BST – and a series of tows at 2 hour intervals through a 24 hour period. Vertical distributions, recorded by the two-stage net, were examined for two resident species, *Atylus swammerdami* (fig. VI.41) and *Bathyporeia pelagica* (fig. VI.7). The former was usually most abundant in the bottom net, but *B. pelagica* seemed to congregate in the surface layers of water at dusk, shortly after it emerged from the sand, and then spread more evenly through the water column as the night advanced (fig. 25). The abundance of night-swimming amphipods varied with the seasons, being low from January to March, increasing in April, decreasing in June and peaking once more in July. Numbers fell again in August, but reached a final peak in September which then tailed off through the autumn (fig. 26). The July peak was not a regular seasonal feature. It consisted mostly of juveniles of essentially subtidal species washed inshore by westerly gales, often in very large numbers. Excluding this phenomenon, the pattern revealed by Fincham's samples illustrated both seasonal cycles of abundance and daily swimming cycles which are displayed by many intertidal, infaunal crustacea.

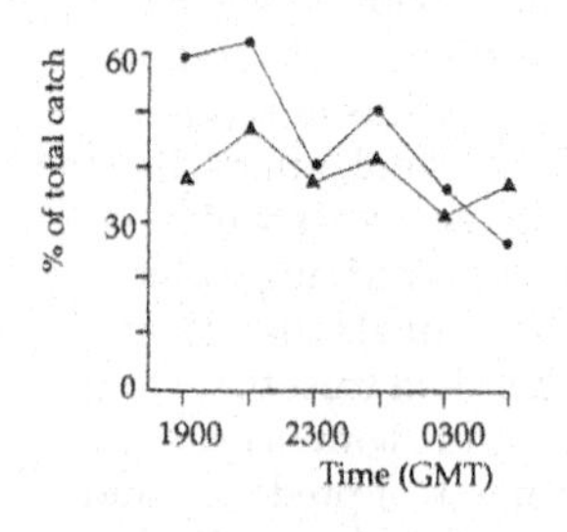

Fig. 25. Proportions of *Bathyporeia pelagica* (dots) and *Atylus swammerdami* (triangles) occurring in the upper of the two surf plankton nets (data from Fincham 1970).

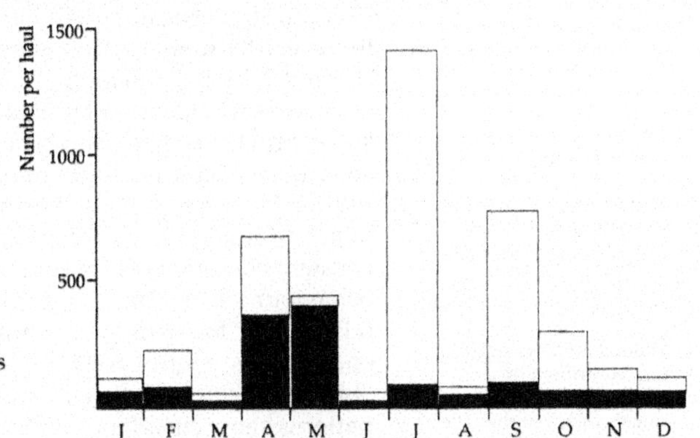

Fig. 26. Average numbers of amphipods in the surf plankton through the year. Resident species shaded; immigrant species unshaded (after Fincham 1970).

The questions posed by these observations are intriguing. Each species of infaunal crustacean occupies a defined zone in the intertidal sands which it must, however, leave at regular intervals. Why and how these small animals are so strictly zoned is not clear. Each must be adapted for life within its optimum habitat; these adaptations perhaps relate to feeding mode, burrowing ability in relation to sediment grade, or tolerance of particular interstitial environmental parameters. Research is still needed to establish the factors determining the distribution of each species, but it has been shown on numerous occasions that each has evolved behavioural patterns which ensure that the animals return to their preferred habitat after each nightly swim. For strictly intertidal species the single, most critical influence on their behavioural patterns is the twice daily tidal cycle. Superimposed on this is the fortnightly spring/neap cycle, which results in the higher levels of the shore being covered by water for only three or four days in every cycle, and the yearly tidal cycle, through which the times of high and low water shift gradually from day to day. In order to maintain their station on the shore, sand dwelling crustaceans thus have continuously to adjust their behavioural cycles to remain in phase with those of the tides. Until quite recently it was believed that their behaviour was controlled by strictly external factors, such as temperature, pressure and the natural light/dark cycle (nL/D). However, it is now known that in most species, while external stimuli may be important in mediating the timing of their swimming behaviour, the activity cycles of these animals result from persistent internal rhythms, under hormonal control.

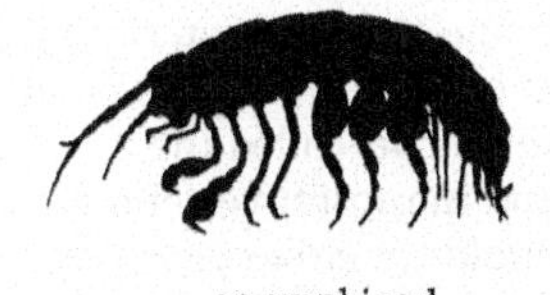

an amphipod

nL/D: the natural cycle of light and darkness

All biological processes – physiological, metabolic and behavioural – show a fundamental daily rhythm related to the rotation of the earth and the resulting natural cycle of light and darkness (nL/D). Thus, successive periods of activity and rest tend to be correlated with this natural rhythm. However, although biological rhythms are related to the earth's daily rhythm, they are not completely controlled by it. In all animals, including intertidal crustaceans, behavioural phenomena such as swimming show a rhythmic pattern which continues even under conditions of constant light or darkness. The period of this rhythm is always slightly more or less than 24 hours, and is hence referred to as circadian, meaning "about a day". If constant conditions are maintained, circadian rhythmic behaviour will continue at its own very precise frequency – it is said to "free-run" – indicating that it has an internal, or endogenous, control system. This control system, or "biological clock", is usually hormonal. Not all behavioural patterns show circadian rhythms; some may be directly stimulated by external, or exogenous, environmental stimuli. They nonetheless display a daily periodicity and are referred to as diel rhythms. With a period of, say, 24.5 hours a free-running circadian rhythm will drift out of phase with the natural L/D cycle, and the animal must constantly reset its internal clock to synchronise its own rhythm with that of the environment. How this complicated system could have evolved throughout the animal kingdom is unknown, but its effect is to allow each organism to fine-tune its own rhythms to the seasonally changing rhythm of the nL/D cycle. Fine-tuning does require an environmental cue, called the *zeitgeber* (literally, "time-giver"), and in most invertebrates this is usually light intensity thresholds at either dawn or dusk. Circadian rhythms are especially complex in motile intertidal animals, which must anticipate tidal times if they are to avoid being stranded at low water. However, as tide times shift daily, the circadian rhythms of marine animals may have superimposed on them additional endogenous rhythms. The period between two high or two low tides is 12.4 hours, and behavioural patterns with this periodicity are termed circatidal. Similarly, the complete spring-neap-spring tide cycle takes 15 days, "about" half of a lunar month, and rhythms with this periodicity are said to be circasemilunar. Biological clocks and circadian rhythms, including those of marine animals, are discussed in depth in Brady (1982).

circadian rhythm: a cyclical rhythm in the metabolism, physiology or behaviour of an animal, with a period of about 24 hr under constant conditions

endogenous (of a biological rhythm): controlled by a physiological process (a "biological clock") within the animal

exogenous (of a biological rhythm): controlled entirely by environmental cues

diel rhythm: a rhythm with a 24 hr period, which occurs under conditions of natural or artificial day/night cycles, but not under constant conditions

zeitgeber: literally "time-giver". A periodic environmental signal that switches on or synchronises a biological rhythm

circatidal rhythm: a biological rhythm with a period of about 12.4 h, equivalent to the period between two high, or low, tides

circasemilunar rhythm: a biological rhythm with a period of about 15 days, equivalent to the period of one tidal cycle

Eurydice pulchra (pl. 2.1) is a predatory isopod common on most sandy beaches. It lives on the upper half of the shore, generally being most abundant between MTL and mean high water of neap tides (MHWN). *Eurydice* leaves the sand as the tide flows in, swimming in search of food, and buries itself in the sand again as the tide ebbs. As the fortnightly spring tides carry water further up the shore, so *Eurydice* moves up, and can be found in the sand right up to the mean high water of spring tides (MHWS). But as the tidal cycle swings again towards neaps, the *Eurydice* population moves down shore again, avoiding being stranded above the neap high water mark (Fish, 1970). Increased pressure, simulating a rising tide, was found to induce *Eurydice* to emerge from the sand and swim upwards (Morgan, 1973), but such a simple response seems unlikely to enable the animal to maintain its position on the shore with reference to the tidal cycle. Experimental animals were uncooperative and would only emerge and swim if the sand at the bottom of the experimental tank was vigorously stirred for up to half an hour, at 12 hour intervals for 4 days. Once this was done they began to show cyclical swimming/resting behaviour, each cycle about the same length as the natural tidal cycle (that is, circatidal), and this continued for up to 10 days after the initial sand-stirring stimulation. Kept without sand, *Eurydice* showed peak swimming activity at 12 hour intervals (fig. 27), again reflecting a circatidal periodicity (Jones and Naylor, 1970).

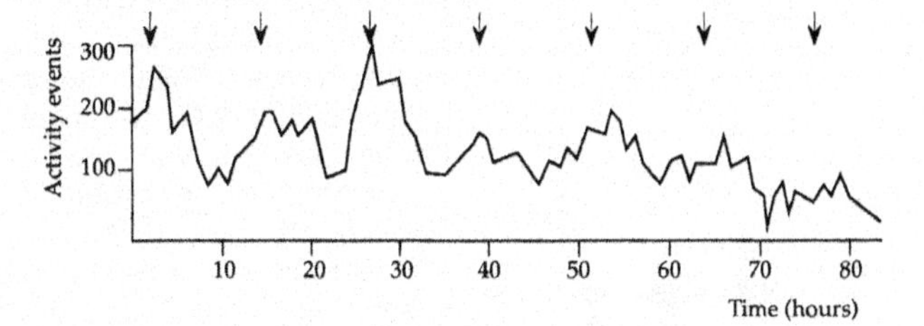

Fig. 27. Tidal swimming of *E. pulchra* under constant conditions without sand. Expected high tides are indicated by arrows (after Jones and Naylor 1970).

Fish and Fish (1972) found that whether or not captive *Eurydice* swam depended entirely on the tidal cycle at the locality from which the animals were collected. During the waxing and waning of the moon, when spring tides are falling, the *Eurydice* population shows peak activity, emerging from the sand and swimming in the surf. During neap tides few animals emerge, and *Eurydice* is usually absent from night time surf plankton. Cyclical swimming could only be induced in experimental animals by stirring their tank during the expected time of high spring tides at their home locality. This rhythmic swimming behaviour in *Eurydice pulchra* has a truly endogenous origin (Reid and Naylor, 1985); it is possible to break this rhythm by chilling animals to 5°C, simulating seawater temperatures in winter, when the population forsakes the intertidal zone and moves offshore. The rhythm may be switched on again by raising seawater temperature and agitating sand in the animals' tank, as described above. Captive animals will then display cyclic swimming activity, with a 12 hour periodicity, which will continue under constant laboratory conditions for nearly 60 days. Initiating this rhythm simply by warming and shaking previously chilled animals is perhaps possible because it reflects natural conditions in early spring, when rising seawater temperature and the much greater movement of water associated with the spring tides of the

March equinox first establish a circatidal swimming rhythm in the *Eurydice* population. In natural conditions *Eurydice* swims mostly at night. Animals emerging from the sand, or washed out by turbulence, during the day show photonegative behaviour and immediately bury themselves again. *Eurydice's* response to light depends on food: hungry animals are photopositive and swim upwards, but as most will feed at night they are sated during daylight hours and thus do not indulge in prolonged daytime swimming. The behaviour of *Eurydice pulchra* is complex, with external factors modulating an endogenous rhythm, which has a period of about 14–15 days, i.e. circasemilunar, corresponding to a complete spring–neap–spring tidal cycle. Under constant conditions the rhythm will run freely for up to 60 days, but as tides change daily it is necessary for the animal to reset its clock at intervals, in order to keep its activity cycles in phase with the tidal cycle. At present, it seems most probable that water turbulence and increased pressure, associated with the spring tides, are the cues which enable *Eurydice* to synchronise its activity rhythm with the tidal cycle.

Species of the amphipod genus *Bathyporeia* (VI.3–VI.9) also leave the sand at night to swim, and like *Eurydice* their swimming shows both a circatidal and circasemilunar periodicity. The animals emerge on the early ebb of high tides, and are two or three times more active on night-time tides than during the day. *B. pilosa*, on some shores, occupies two intertidal zones (fig. 28), which do not shift in response to the spring/neap cycle, but remain constant even though the high shore population must have less opportunity for swimming than those on the lower shore. However, all *Bathyporeia* species feed within the sediments, licking food items from individual sand grains, and their swimming activity is related to their reproductive cycles (Preece, 1971) rather than to a search for food. Peak activity occurs around the periods of the spring tides, coinciding with the 15-day cycle of embryo development in reproducing females. *Bathyporeia*'s endogenous rhythm is modulated in the same manner as those of *Eurydice*, with temperature, nL/D and tides acting as exogenous synchronizing factors. As in *Eurydice*, the endogenous rhythms of *Bathyporeia* will free-run in animals kept under constant experimental conditions: maximum swimming activity occurs at predicted night-time high tides, and reversing the experimental light/dark cycle will reverse the activity peak of captive animals, showing that light is an important *zeitgeber*, switching on the behavioural cycle (fig. 29). However, it is not yet known which exogenous stimulus is most important in re-phasing the activity cycle to keep in tune with seasonally changing tidal cycles and nL/D ratios.

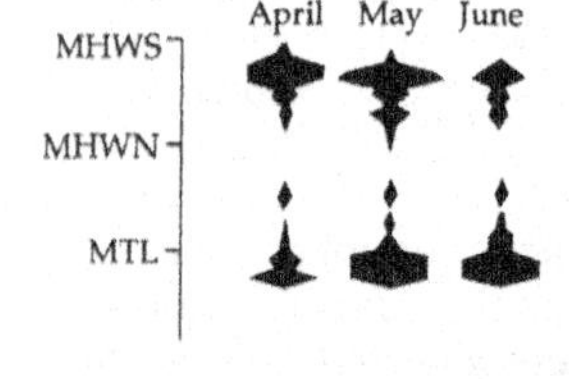

Fig. 28. Distribution of two populations of *Bathyporeia pilosa* in relation to tidal level (after Fish and Preece 1970).

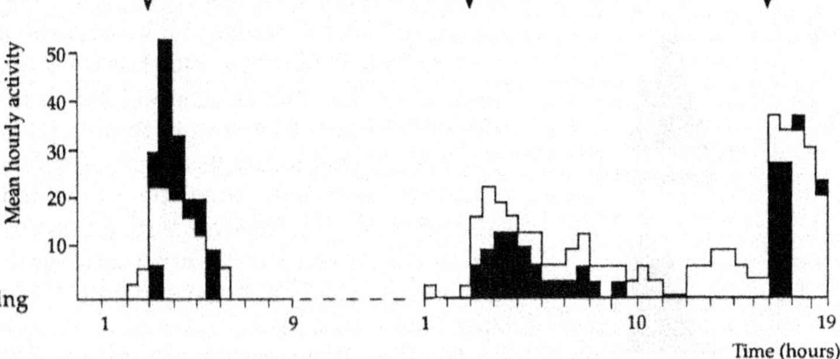

Fig. 29. Tidal swimming activity in *Bathyporeia pilosa*. Shaded, under natural conditions; unshaded, in continuous dark. High tide times indicated by arrows (after Preece 1971).

supralittoral: that part of the beach environment above the level of the highest tides

Most interesting of all are the behaviour patterns shown by the supralittoral sandhoppers, amphipods which spend their entire lives above high water mark. *Talitrus saltator* (pl. 2.4) is the most familiar of these. *Talitrus* spends the daylight hours buried in the sand beneath strandline debris, but emerges at night to range down shore in search of food. Like Cinderella it must be home before dawn, and ahead of the incoming tide. The animal's emergence and activity cannot be stimulated by changes in light intensity, nor water pressure or turbulence, and in order to return safely above high water mark it must be capable of orientated locomotion. Consequently it is not surprising that the activity rhythm of *T. saltator* is under precise endogenous control, and in constant conditions will free-run for more than 100 days without variation (Williams, 1983). The endogenous activity rhythm of *Talitrus* has a circasemilunar periodicity, with activity greatest on falling spring tides (Williams, 1979); through this period the zone occupied by the animals moves down shore with the strandline, to between high water spring and high water neap tide marks, and upshore again as the spring tides approach. This movement is related to exogenous factors, especially humidity, which must be above about 90% for resting animals to survive. *Talitrus* has no endogenous circatidal element in its daily activity cycle, which, as in other supralittoral beach animals, is entirely circadian. Its daily activity occupies a six to eight hour period of the night, with the peak always occurring between 0100 and 0300 hours GMT regardless of the state of the tide. This circadian rhythm has a period of 24.46 hours and when free-running a perceptible daily shift is apparent (fig. 30), which must be constantly rephased in relation to the seasonally changing nL/D ratio. Light is the obvious synchronizing cue for a nocturnal supralittoral animal, and in *T. saltator* the dawn transition marks an immediate change in its behaviour. One hour before dawn a light intensity threshold

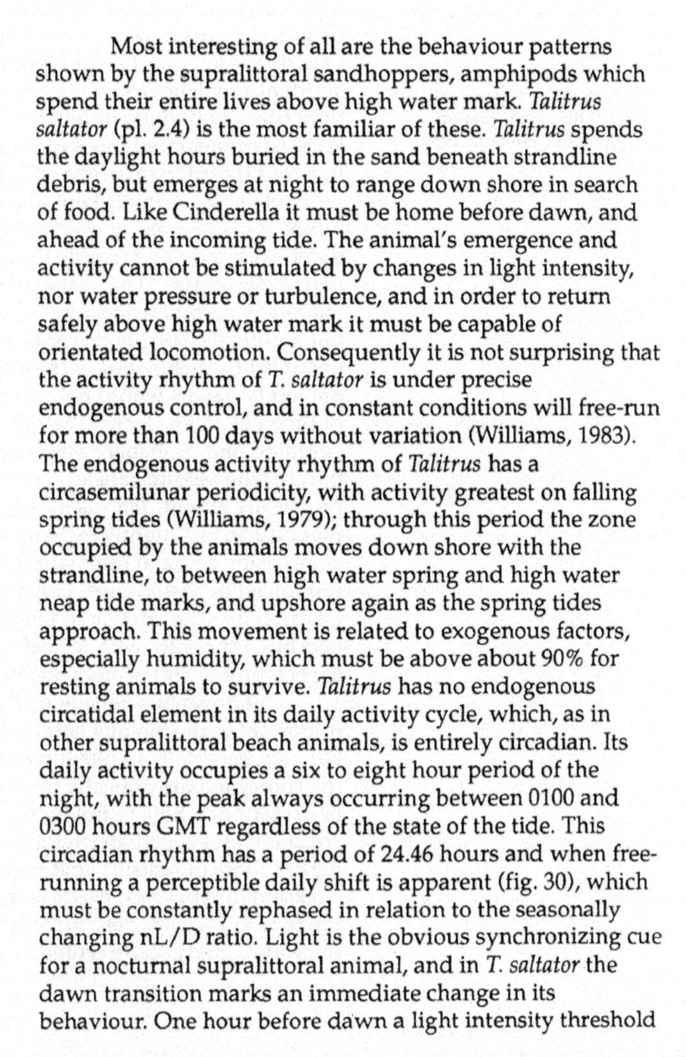

Fig. 30. Free-running activity rhythm of *Talitrus saltator* under constant light over a nine-day period (after Williams, 1983).

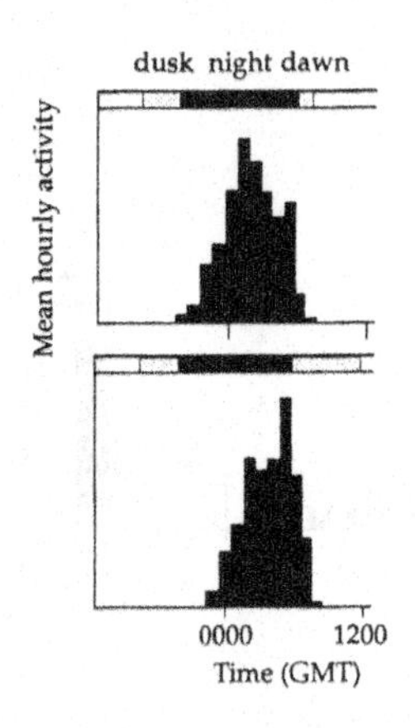

Fig. 31. *Talitrus saltator* activity rhythm under experimental L/D, with variable dawn length (after Williams 1980).

of about 1.5 lux signals an end to foraging, and the amphipods abruptly cease random movement and head upshore. Maximum moonlight intensity on British beaches ranges from 0.3–0.8 lux and thus does not influence the animals' behaviour. Williams (1980) found that it is the dawn light transition, rather than the dusk, which allows *Talitrus* to synchronise its activity periods with the nL/D cycle. At dawn light intensities of less than 1.5 lux the circadian rhythm free-ran, with a perceptible drift, but with a dawn light value of 1.5 lux or greater the activity rhythm achieved a steady relationship with that of the experimental nL/D ratio (fig. 31).

The dawn upshore migration of *Talitrus* is also controlled by an endogenous circadian rhythm, which may be independent of that controlling emergence and foraging. During the later part of its nightly activity, *Talitrus* quite suddenly switches to orientated movement towards a light/dark boundary, or horizon. Edwards and Naylor (1987) demonstrated this experimentally, observing animals in a round cardboard arena with part of its perimeter painted black. Animals kept in continuous dark and then released in the arena at different times all behaved identically, moving towards the dark horizon exactly one hour before dawn. The timing of this response could be altered by acclimatising animals to different nL/D ratios, but a difference of 2 hours or less between two regimes did not affect the behaviour of the animals, suggesting that dull or cloudy dawns would not disturb the normal functioning of their circadian clocks. The behaviour of *Talitrus saltator* is complex, and probably varies between populations. On parts of the Mediterranean coasts of Europe, where tides are practically absent, *Talitrus* moves inshore beyond high water mark during its nightly expeditions, and navigates home again using celestial orientation, with a circadian timing, reinforced by visual cues. There is scope for much more research into the orientated behaviour of *Talitrus* on British beaches, especially with regard to local variations related to aspect, profile and geomorphology of different beaches.

Endogenous behaviour rhythms are thus seen to be important in the biology and ecology of motile, intertidal animals, especially in relation to their zoned distribution on the shore, and are critically important in synchronizing the behaviour of whole populations. The varying significance of circatidal and circadian rhythms between lower shore and higher shore organisms, and their relationship to circasemilunar cycles, needs to be investigated in a wider range of animal species, but is apparently fundamental to intertidal, sandy shore ecology.

5 Populations, competition and predation

Populations of infaunal sandy shore animals show different structures from species to species. In some, recognisably distinct age groups may be charted through successive years, from their initial recruitment to the population as juveniles, to their eventual disappearance on reaching maximum age for the species. In others, populations tend to be dominated by a single age class, and often show a recurring pattern of boom and bust, sometimes punctuated by complete local extinction, after which the species may be entirely absent from a beach for years at a time. While competition and predation appear to be important in modulating populations of, especially, species with the latter type of population cycle, they perhaps exert a less significant effect on species with more stable cycles.

Many bivalve molluscs characteristically show irregular fluctuations in numbers. Their reproductive behaviour involves broadcasting eggs and sperm into the water, where fertilisation occurs, and the development of a planktonic larval stage. Reproductive success is usually improved by synchronous spawning within each population, but most of the resulting larvae are dispersed by the tide. There is some value in this in that filter-feeding bivalves do not distinguish between food particles and their own young, and in that the sand meiofauna includes a large number of small predators. Thus, it is perhaps only through their involuntary dispersal away from their origin that the larvae achieve a reasonable rate of survival. Bivalves may be aged by counting the annual growth rings on their shells (fig. 32). Each ring represents a winter check, when growth temporarily ceases, and the bands between each ring represent periods of rapid summer growth. The number of annual rings correlates with other parameters, including the

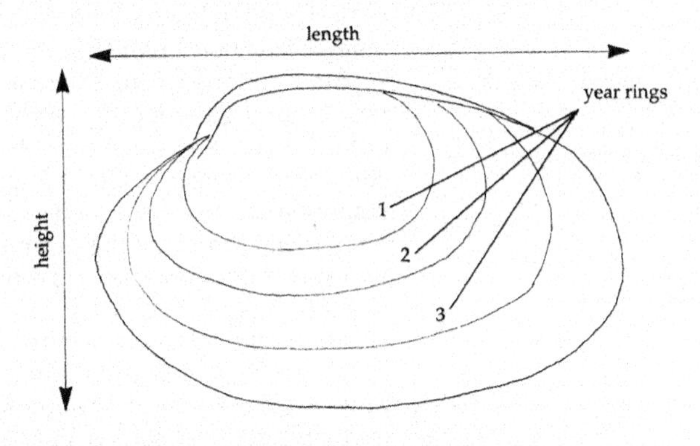

Fig. 32. Measuring and ageing a bivalve shell.

length, height and breadth of the shell, and shell and body weight (fig. 32). The correlation may be isometric, with all parameters increasing at the same rate, or allometric, as in the Cockle, in which the shell becomes relatively broader and heavier with age. In most species the first few years' rings are usually quite obvious, but in some the later rings become harder to distinguish from shell banding caused by other environmental parameters, and so year groups tend to merge as they age. In these cases length/frequency plots will give some information on the age structure of the population. When surveying the infauna of a particular beach, additional interesting data can be collected by measuring length and counting rings on all specimens of each species of bivalve found. This is easily done in the field using a vernier scale, or even more simply with graph paper glued to a board.

The weight of flesh contained within the bivalve shell also increases with the age of each animal, but varies seasonally according to the cycles of growth and reproduction of each population. Most bivalves on British beaches show similar annual cycles: winter is a period of inactivity during which flesh weight of the animal declines to a minimum in February or March; through this period the bivalve maintains a low metabolic rate, and survives by drawing on stored reserves of glycogen. Growth resumes in April with the onset of the rich spring phytoplankton bloom, and body weight increases rapidly as the bivalve rebuilds its reserves and uses surplus energy to fuel the development of eggs and sperm. Spawning begins in early summer, when the seawater temperature exceeds a critical value, and in all British intertidal species is quite protracted, in some cases extending well into September. *Chamelea gallina* (formerly *Venus striatula*) (pl. 4.9) breeds between April and October, but spawning is concentrated in two periods – April to July, and August to October – and there are thus two main periods of recruitment of juveniles to the population (Guillou and Sauriau, 1985). The first spawning period is often longer than the second, and length/frequency plots (fig. 33) consequently show three separate cohorts within the first year group; however, in year 2 rapid growth of the whole population tends to iron out differences between the cohorts, and by year 3 only a single peak is visible. Growth rate in bivalve populations responds sensitively to food availability, and fluctuations in body weight through the summer reflect the interaction of two synchronised cycles: reproduction, and the storage and utilisation of reserves. In populations of the Wedge Shell, *Donax vittatus*, at Gullane in the Firth of Forth (Ansell and Bodoy, 1979), increase in tissue weight during April and May was due to rapid development of the gonad, which by mid-summer represented 35–40% of total tissue weight. This

cohort: a group of individuals arising from a single reproductive period or event. A year class may consist of several cohorts.

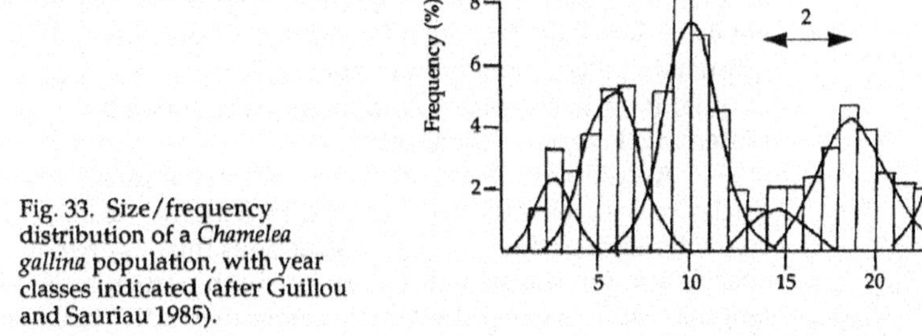

Fig. 33. Size/frequency distribution of a *Chamelea gallina* population, with year classes indicated (after Guillou and Sauriau 1985).

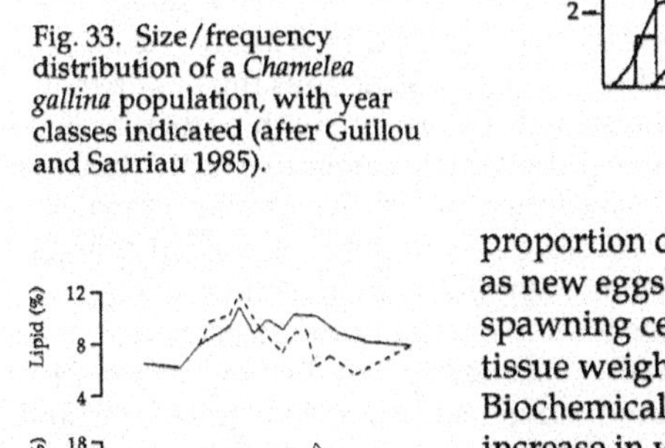

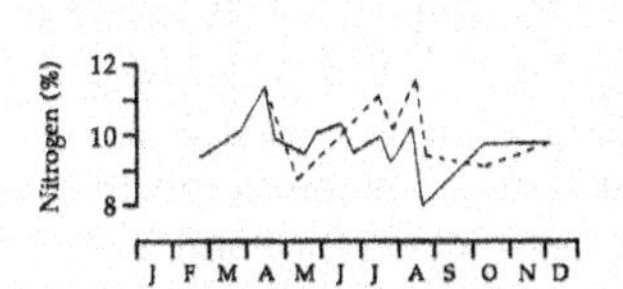

Fig. 34. Changes in biochemical composition of *Donax vittatus* through one year (after Ansell and Bodoy 1979). Solid line denotes males, dashed line females. A single line indicates sex could not be determined.

proportion declined at each spawning, and increased again as new eggs and sperm were produced. In August spawning ceased, the remaining gonad was resorbed, and tissue weight began to decline towards its winter minimum. Biochemical analysis of *Donax* tissue showed that the spring increase in weight included a disproportionate increase in the amounts of lipid and carbohydrate (fig. 34), both of which fluctuated through the summer, reflecting successive spawnings. Following the end of spawning lipid levels declined only slowly but carbohydrates, which constitute the main energy reserves of the animal, dropped sharply through the winter as they were drawn upon to supply its metabolic needs. The amount of energy expended in reproduction by bivalve populations may be considerable: in 5-year old *Angulus tenuis* at Loch Etive (Ansell and Trevalliou, 1967) it accounted for 18.2% of the total annual weight increase of the population (fig. 35).

These annual cycles of growth and reproduction naturally vary between different populations of each species, and are profoundly influenced by food availability. Cockles generally need to be covered by water for at least five hours in each tidal cycle in order to feed enough to support normal growth rates. In high shore populations studied by Jones (1979), the cockles were covered for only 3–3 3/4 hours in each cycle and as a result grew more slowly and achieved a smaller maximum size than animals from low shore populations. Intertidal populations of bivalves are frequently dominated by one or few year classes, indicating that the recruitment and survival of juveniles vary from year to year. A good example is that of *Donax vittatus* in Kames Bay, Millport, which achieved successful settlement in only two years between 1926 and 1946 (Ansell, 1972). In 1962 another successful spatfall occurred which by 1965 gave a population estimated at 3.87 million individuals, with a mean density of 73 m^{-2}. By 1967 natural mortality had reduced this to 0.87 million, with a density of 16.5 m^{-2}, and by 1969 the population was so low that no living specimens could be found at LWST. During 1967 and 1968 *Donax* had

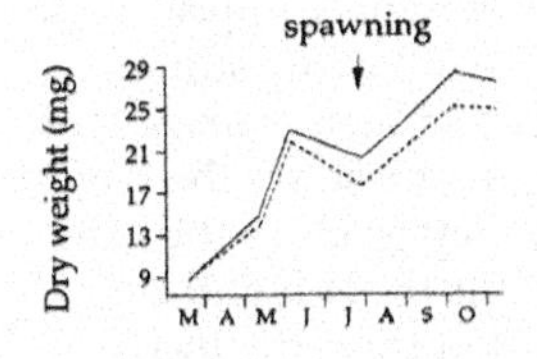

Fig. 35. *Angulus tenuis*: changes in total weight (solid line) and soft tissue weight (dotted line) through part of 1966, showing decrease in soft tissue at spawning (after Ansell and Trevalliou 1967).

displayed its usual cycles of growth and reproduction, and released large numbers of eggs and sperm. However, larval survival and settlement were evidently badly affected by as yet unknown environmental factors and the population ultimately failed. Equally dramatic fluctuations may occur in other long-lived species, and might be revealed by systematic long-term monitoring.

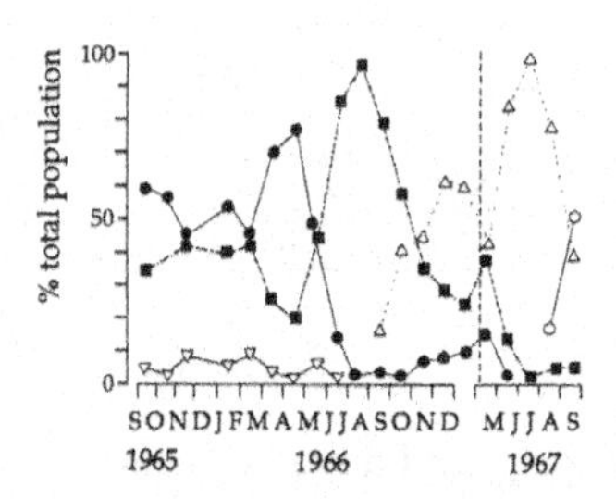

Fig. 36. The population structure of *Eurydice pulchra*, showing three separate components at any one time. ▽ = oldest animals, ■ = juveniles, ● = mature adults, △ and ○ = new juvenile generations. (After Fish 1970.)

Amphipods and isopods, like all crustaceans, grow by discrete increments, each marked by a moult. Adult characteristics and reproductive structures thus appear at specific growth stages, which correlate with particular body lengths, and it is generally easy to categorise individuals as, for example, juveniles, immature adults, or mature males and females. It is therefore easy to trace individual cohorts through a succession of length/frequency plots, even though generations tend to overlap through each year (fig. 36). These small crustaceans rarely live more than 24 months. Their reproductive seasons may be quite long, and are usually found to encompass one or two, rarely three, peak periods of breeding activity. As in all crustaceans, mating, and the release of juveniles, are synchronised with the moult cycle, which is itself, in intertidal species, related to the tidal cycle. *Eurydice pulchra* populations may occur at densities of 1500 m^{-2} or more, and on a South Wales beach studied by Jones (1970) exceeded 7380 m^{-2}. This species breeds between April and August once sea temperatures pass above 10°C, and the highest numbers of juveniles occur around the periods of maximum summer temperatures (fig. 37). Males and females pair during their nightly swimming on falling spring tides and mating occurs in the sand once the female has completed her moult. Mating takes place following each spring tide from early April and the females thus produce a sequence of broods through the summer. Juvenile *Eurydice*, 2–2.5 mm long, first appear in June and mature through the winter to begin breeding the following spring. They die off in the autumn at 15 months of age. Mid-summer juveniles also mature to breed the following summer and reach only 12 months of age before dying. However, the last broods of juveniles, appearing in October, do not mature until late the following summer, breed once in their second October and then overwinter for a second time. In the following spring these are the largest individuals in the population and they produce the early broods of the year, before dying off at 18–24 months old. *Eurydice pulchra* populations consist of three distinct, overlapping cohorts, except in the period from June to September, by when the oldest individuals have all died off.

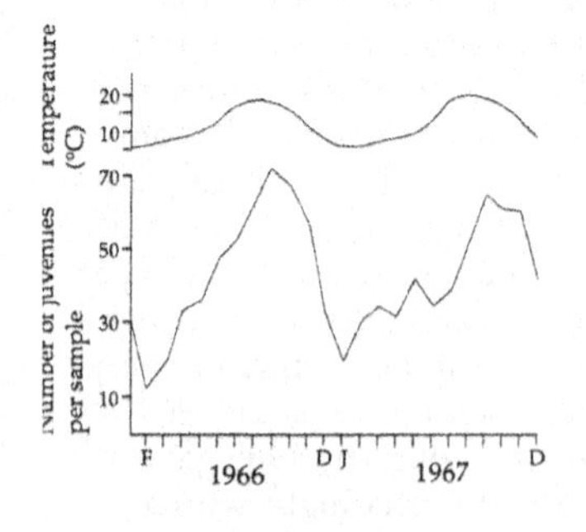

Fig. 37. *Eurydice pulchra*: numbers of juveniles in monthly samples in relation to mean sea temperature (after Jones 1970).

The amphipod *Bathyporeia pilosa* is abundant on most moderately exposed to sheltered sandy shores. Fish and Preece (1970) recorded maximum densities of more than 3000 m^{-2} on a sheltered West Wales beach, with females

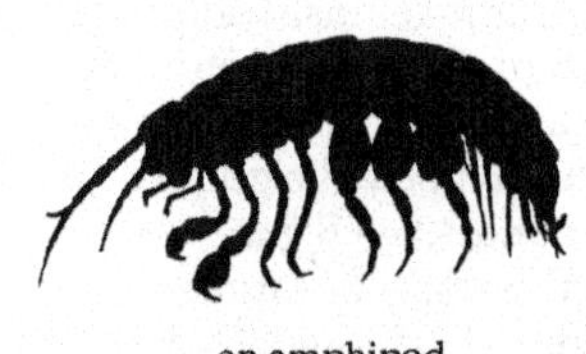

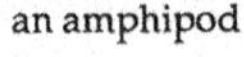

an amphipod

an isopod

constituting the bulk of the population throughout the year. On this shore *B. pilosa* reproduced in all months of the year, with one-third of adult females bearing eggs even in the winter months. The greatest reproductive activity, however, still occurred during summer, with pronounced peaks in April–May and August–October. The reproductive cycle takes just 15 days to complete, from fertilised egg to free-swimming juvenile. Adults pair and mate in night-time ebb tides following each new and full moon, after which juveniles are released and the female moults and mates again. The overwintering population of *B. pilosa* consists largely of juvenile animals. These mature in spring to form the majority of the next breeding population and eventually die in June and July. Their offspring mature swiftly and breed in the autumn; a proportion will survive the winter, as mature breeding adults, but most die, while their offspring provide the new generation of breeding adults.

The breeding cycle of the supralittoral sandhopper, *Talitrus saltator*, forms an interesting contrast to those of the intertidal amphipods and isopods. This species has a shorter breeding period, from May to the end of August, which is controlled by day length rather than temperature. *Talitrus* breeds when the natural day length reaches 16 hours, irrespective of air and sea temperature (Williams, 1978). The population comprises two cohorts through most of the year. Juveniles are numerous in June and July (fig. 38) and usually mature by the autumn but do not breed until the following summer. The overwintering population consists of these young adults, together with a small number of juveniles representing the last brood of the season, and a proportion of large sexually mature males and females, which represent the last of the breeding population of the year. In February, these large old animals die off and the young adults, with the now maturing juveniles, form the new breeding population.

Intertidal amphipods are ideal subjects for population studies on sandy shores. They are often abundant and relatively easily sexed, and can usually be sorted into age classes simply by measuring length. Comparatively few species have been studied in detail, and on almost any moderately sheltered shore regular sampling through one complete yearly cycle will show interesting variations in population density, sex ratio and age distribution.

Populations of the Lugworm, *Arenicola marina*, are limited by food (Farke and Berghuis, 1979). Both the weight of individual worms and the total biomass of the population fluctuate considerably through the year, and growth rates tend to be irregular. There is no accurate way of ageing Lugworms beyond the 6 mm-long post larval stage, called the Benham larva (fig. 23). The smallest, palest individuals

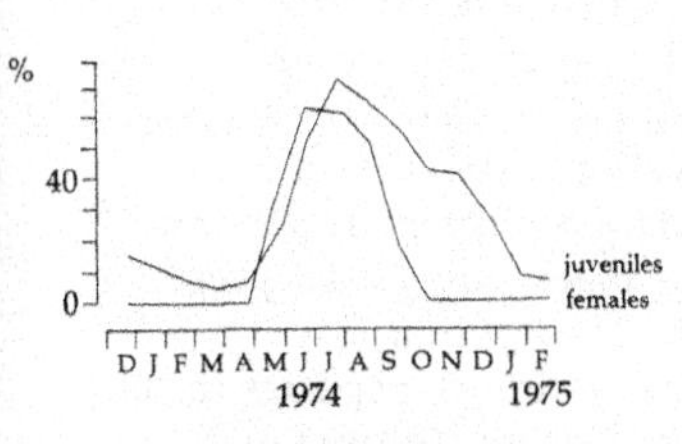

Fig. 38. *Talitrus saltator:* numbers of juveniles as % of total population through the year, with numbers of egg-bearing females as % of total females (after Williams 1978).

in a population obviously represent the latest recruits, but subsequent growth depends so much on available food that plots of weight against frequency or length against weight do not give an accurate representation of the age structure of the population (De Wilde and Berghuis, 1979). In the sandflats of the Dutch Wadden Sea, mean body weight of *Arenicola* averaged 2–2.5 g in summer and 4.5–5.5 g in winter; few individuals reached 6 g, although the largest, and presumably most well-fed, individual weighed 9.5 g. Like many northern, cold water species, *Arenicola* breeds late in the year and, in all populations so far studied, has a protracted spawning period between September and November, consisting of two peaks, in late September and late November, with a five-week period between during which no spawning occurs. In February and March active Lugworms begin to produce eggs and sperm, which accumulate in the body cavity of the worm. By July the entire population consists of sexually mature males and females, usually in equal numbers, which are easily distinguished by viewing their gametes. Spawning is synchronous, induced by falling seawater temperatures at a threshold of 13°C, or by an abrupt downward temperature shock (Farke and Berghuis, 1979). Males spawn first, discharging sperm into their burrows, from which it is ejected by the pumping action of the worm to accumulate in pools on the surface of the sand. The females discharge eggs into their burrows, where they are fertilised by sperm drawn in as the female ventilates its burrow. The eggs develop and hatch in the female Lugworm's burrow, the larvae leaving it for their upshore migration when they have grown just three chaetae-bearing body segments. Neither male nor female Lugworms feed during their spawning, and the females do not begin to feed again until well after the larvae have left the burrow, behaviour which almost certainly reduces the risk of eggs, sperm and larvae being engulfed and eaten by the adult worms. Why *A. marina* should have two distinct periods of spawning is not clear. It is possible that the November and September spawners represent genetically distinct components of the population, although this has not been tested by geneticists. From several studies it is clear that the two periods of spawning activity occur at the same tidal levels and there is no evidence that they result from two topographically separate populations. It is known that large, black Lugworms present low on the shore on some beaches – the laminarian Lugworm or "Black Lug" – is in fact a different species from the mid-lower shore *A. marina*, the so-called "Blow Lug" (Cadman and Nelson-Smith, 1993), but there is as yet no evidence that these spawn at different times.

Competition and predation doubtless play important roles in modulating the population dynamics of intertidal infaunal animals, but with a few well-studied exceptions there is at present little quantitative information on these topics. Competitive interactions seem especially difficult to demonstrate in an extensive, and apparently uniform, habitat such as a sandy beach, but it is probable that, at least in some bivalve species, population structures are partly modulated by intraspecific competition. For example, natural populations of the Cockle, *Cerastoderma edule*, may be dense enough to occupy 40% of the surface area of the beach in their preferred zones (Jensen, 1985). It is known that population densities of the small Tellin *Macoma balthica* (pl. 4.8) may be depressed by high populations of Cockles, but one reason why cockle beds are usually dominated by a single year class is that at high densities they exert a negative effect on the settlement and survival of their own juveniles. Cockle beds are periodically decimated by severe winter weather, and a high winter mortality is often followed by an exceptionally heavy spring spatfall. The post larval Cockles then grow rapidly to occupy the maximum surface area of sand within a year. Thereafter, new generations of post larval Cockles may be prevented from settling by the exhalant currents of the resident adults, or on achieving settlement may be crushed by the adults' regular movements in the sand. A small, burrowing amphipod, *Corophium volutator* (pl. 2.5), is excluded from areas dominated by Cockles by similar effects. *Corophium volutator* is common in silty areas of sand generally unsuitable for Cockles. However, Jensen (1985) found that following a winter mortality of Cockles, *Corophium* moved into sandy low-shore regions where it normally did not occur. As the Cockle population was renewed, *Corophium* declined; the constant movement of the Cockles destroyed the amphipods' burrows, preventing them from feeding and eventually excluding them from large areas of the shore. Much remains to be discovered about competitive interactions among species of sand-dwelling animals. Good results could be obtained from careful field observation, combined with regular monitoring of population numbers and experimental investigation of behaviour.

Hunting predators of sandy shores include the Green Shore Crab, *Carcinus maenas* (pl. 5.3), the specialized prosobranch snail, *Lunatia*, several opisthobranch sea slugs, and a number of carnivorous bristle worms, especially *Nephtys hombergi*. *Carcinus* is not a permanent resident of intertidal sands, but will range along the shore, from adjacent rocks, mussel beds or mud banks, feeding where opportunities allow. The natural population of Cockles studied by Jensen (1985) had an initial settlement density of 40,000 spat m^{-2}, which declined to 1200 m^{-2} in 18 months, almost entirely through predation by Shore Crabs. Beyond 18 months of age the Cockles were said to have achieved a

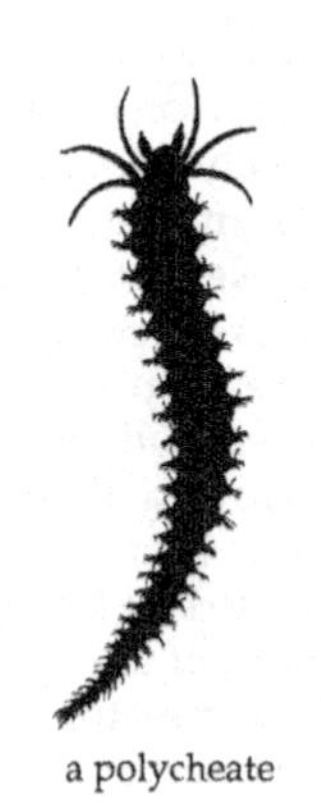
a polycheate

"size refuge", with shells sufficiently large and thick to resist *Carcinus* predation. The Necklace Shells *Lunatia catena* and *L. alderi* are also efficient predators of bivalves, drilling through their shells close to the umbo (fig. VIII.1). *Angulus tenuis* is among the preferred prey of Necklace Shells, and a high proportion of empty *Angulus* valves found along a strand line will bear the characteristic round hole made by the predators in one of the valves. At Kames Bay, Millport, predation by *Lunatia alderi* accounted for 15% of the population of first year *Chamelea gallina*, 5% of second year and 2% of third year animals (Ansell, 1961); by the time the surviving *Chamelea* entered their fourth year, they too had probably achieved a size refuge from *Lunatia* predation.

Nephtys hombergi feeds particularly on other species of infaunal polychaete. Beukema (1987) studied populations of *N. hombergi* and its two principal prey species, *Scoloplos armiger* and *Heteromastis filiformis*, over an 18 year period in the Wadden Sea sand flats. Numbers of *N. hombergi* fluctuated widely through this period, as the population was severely reduced by prolonged cold winters. Populations of the two prey species increased dramatically in the absence of *Nephtys* and dropped sharply once it increased again. However, the mean biomass of *Heteromastis*, which is swallowed whole by *Nephtys*, decreased more markedly than that of *Scoloplos*, which only has its tail bitten off by the predator.

The primitive, shelled, tectibranch sea slug, *Acteon tornatilis*, is a specialist preying on the tube worms *Lanice conchilega* and *Owenia fusiformis*. *Acteon* is most easily found on a rising tide, when its deep trails may be seen meandering over the sand surface. The animals may travel for 10 m or more, probably navigating by as yet unknown means in order to maintain their preferred position on the lower shore where their prey species occur. On finding a *Lanice* tube, which protrudes an inch or so above the surface, *Acteon* immediately burrows, following the tube to its end, 10 cm or so below, where it seizes the tail of the worm and proceeds slowly to eat it. *Acteon* may take up to two hours to consume an average sized *Lanice* (Yonow, 1989).

The major marine predators of the sand infauna are fish, especially the Sand Goby (*Pomatoschistus minutus* (Pallas)), Sole (*Solea solea* (L.)) and Plaice (*Pleuronectes platessa* L.). Gobies feed mainly on polychaete worms and are permanent members of the sand fauna, reaching maximum abundance between April and September. Le Mao (1986) reported Goby densities equivalent to 190 per 100 m^2 in the Rance Estuary, France. Sole move across sheltered, slightly silty beaches at night, feeding on polychaetes and especially small bivalve molluscs. However, juvenile Plaice are the most significant fish predators, and sheltered sand bays are important nursery grounds for first year, or O-group, Plaice, which feed initially on small polychaetes and bivalve spat, and as they grow, on the protruding siphons of *Cerastoderma edule*, and *Angulus tenuis*. The long, extensible siphons of *Angulus*, in particular, are an important food for young Plaice and, to a degree, are a renewable resource in that the bivalve will actually regenerate its siphons after they have been bitten off by feeding Plaice. Edwards and others (1970) studied captive populations of first year Plaice in large aquaria seeded with different densities of *Angulus* and found that growth rate of the Plaice was dependent on the ratio of predator to prey. Initially, the young fish fed well, cropping unpredated *Angulus* siphons, but the feeding rate and growth of the fish thereafter depended on the rate of siphon regeneration by *Angulus*, and was modulated by competition between the fish. Cropped *Angulus* regenerated their siphons in about three weeks, a growth rate equivalent, per *Angulus*, to 0.04 mg dry wt day^{-1}, and thus a natural population of, for example 200 m^{-2} would yield a daily production of 8 mg dry wt m^{-2} day^{-1}. *Arenicola* is also cropped by flatfish, most especially by group I and II (second and third year) Plaice, and by Flounder (*Platichthys flesus* (L.)). The Lugworm extends its tail from its burrow several times daily in order to defaecate, and the distal tail segments may then be bitten off by feeding flatfish. In the Dutch Wadden Sea (De Vlas, 1979), 50% of the stomach contents in every I and II group Plaice examined consisted of *Arenicola* tails, and 90% of the wild population of *Arenicola* had lost tail segments. Each worm lost 20% of its tail segments annually, representing as much as 110 mg dry wt of tissue, and the total weight of *Arenicola* cropped each year was calculated at 1.6–3.6 gm dry wt m^{-2}, equivalent to 23% of the total *Arenicola* biomass. Unlike *Angulus*, *Arenicola* does not regenerate its tail. It can restore tail length through growth of the remaining segments, but continued cropping ultimately proves fatal to the worm. In the population studied by De Vlas (1979) mean tail size was reduced from about 100 segments to less than five in about six years, and thus the life span of *Arenicola* was limited to five or six years by fish predation.

WORMS OR WORM-LIKE ANIMALS	
BODY SEGMENTED	Phylum Annelida Class Polychaeta **C. Bristle worms** Body divided into ring-like segments, each with paired bundles of bristles on each side, sometimes small but always present. Head usually obvious, with antennae, tentacles, palps and eyes. Flat, soft scales and tentacle-like cirri often on upper side of body.
BODY NOT SEGMENTED	Phylum Nemertea **B. Ribbon worms** Flat, delicate worms, often brightly coloured. No bristles, antennae or tentacles. Head indistinct. Fig. 40.
	Phylum Sipuncula **D. Sipunculans** Fat, cylindrical animals. No head, bristles or antennae. Tapered towards one end, from which a proboscis is extruded, with the mouth at its end.
	Phylum Hemichordata Class Enteropneusta **P. Acorn worms** Long, floppy worms, often smelling of iodine. Head end with a rounded or cylindrical proboscis encircled by a thick collar, thus resembling an acorn.
	Phylum Echinodermata Class Holothurioidea ***Leptosynapta inhaerens*** Pink, sticky, slightly transparent. Mouth at one end, surrounded by 12 branched tentacles.
	Phylum Cnidaria Class Anthozoa **A. Sea anemones** Elongate, cylindrical, contracting to an irregular sphere. Showing a flat disc of tentacles surrounding a terminal mouth when relaxed.

NOT WORMS: WITH JOINTED LEGS	
LEGS ALL SIMILAR	Phylum Crustacea Class Malacostraca Order Isopoda **H. Isopods** Body flattened from top to bottom. With short head, segmented middle region and distinct tail portion.
	Phylum Crustacea Class Malacostraca Order Mysidacea **G. Mysids** Shrimp-like animals with large, stalked eyes. Head and middle region covered by a carapace. Legs two-branched, feathery.
	Phylum Crustacea Class Malacostraca Order Cumacea **F. Cumaceans** Tadpole-shaped, eyes indistinct. Carapace swollen; hind portion slender, with forked 'tail'.
	Phylum Crustacea Class Copepoda Order Harpacticoida **E. Harpacticoid copepods** Body long and tapering. Head with one pair of short antennae. Very small legs at front end; 'tail' forked.
LEGS DISSIMILAR	Phylum Crustacea. Class Malacostraca Order Decapoda Suborder Pleocyemata (Reptantia) **K. Crabs** Body flattened. Upper surface covered by carapace, abdomen folded beneath. First pair of legs modified as claws.
	Phylum Crustacea. Class Malacostraca Order Decapoda Suborder Dendrobranchiata (Natantia) **J. Shrimps and prawns**. With long abdomen and distinct tail fan projecting behind carapace. Several pairs of legs may have pincers.
	Phylum Crustacea. Class Malacostraca Order Amphipoda **I. Amphipods – Sandhoppers** Body flattened from side to side. With head, segmented middle region bearing legs, and distinct tail portion. First two pairs of legs with claws or pincers.

NOT WORMS: WITHOUT LEGS	
WITH A HARD OUTER SHELL	Phylum Mollusca Class Bivalvia **L. Mussels, cockles and allies** Shell consisting of two valves hinged along dorsal edge. Head absent, foot and gills well developed.
	Phylum Mollusca Class Gastropoda. Subclass Prosobranchia **M. Sea snails: prosobranch gastropods** Shell of one piece only, cylindrical, globular or spiralled. Head present, with tentacles.
	Phylum Mollusca Class Gastropoda. Subclass Opisthobranchia ***Acteon tornatilis*** Shell spiralled, banded pink and white. Head lobed, without tentacles. Shelled sea slug.
WITHOUT AN OBVIOUS SHELL	Phylum Mollusca Class Gastropoda. Subclass Opisthobranchia **N. Sea slugs** Body soft, slug-like, sometimes with small internal shell. Head end lobed or with short tentacle-like projections.
	Phylum Echinodermata Class Echinoidea **Oi. Heart urchins** Hard, heart-shaped, covered with bristles when living. Upper surface with petal-shaped grooves.
	Phylum Echinodermata Class Asteroidea **Oii Starfish,** with five or more short arms Class Ophiuroidea **Oiii Brittle stars**, with five long thin arms
	Phylum Annelida Class Polychaeta ***Aphrodita aculeata*: Sea Mouse** Oval, up to 200mm long. Upper surface with thick mass of bristles and iridescent spines. Lower surface segmented.

Fig. 39. A quick-check key to the main groups of invertebrates of sandy shores.

6 Identification

Many common and familiar animals can be picked up on sandy shores and require only a brief reference to a field guide to be easily identified. Digging, sieving and netting reveal a far richer fauna than might be expected. While many bivalves, crabs and shrimps, and some of the larger polychaete worms, are also striking enough to appear in popular seashore guides, specialist literature has generally been required to identify most of the smaller worms and crustaceans. Identification of the sand macrofauna need not be difficult. Most of the animal types encountered will be readily recognized, and for several groups – most of the molluscs, for example – the commonest species may be identified simply by referring to a good illustration. Sampling and identification of the meiofauna require special techniques and a specialist literature, and meiofaunal animals are not covered in the following keys.

All organisms collected should initially be examined alive, preferably on the shore, in white plastic trays of seawater. Specimens which can be identified accurately in the field may then be returned to their appropriate beach level. With foresight it is quite possible to collect much of the information one might require, for instance on numbers, weights, and lengths, without removing animals from the shore. Specimens collected for further examination, or laboratory experimentation, need to be transported carefully, and preferably stored outdoors in shaded, aerated seawater tanks. Polychaete worms are often especially delicate, fragmenting readily on collection. This makes identification more difficult. In most cases it will be necessary to relax the worm with magnesium chloride (technique, p.93) in order to examine fine morphological details, and for some it is important to evert the proboscis by pressing the specimen gently with a seeker handle behind its head. Most amphipods, cumaceans and other small crustaceans can be identified only by microscopical examination, after killing the specimen with 70% alcohol. A hand lens is helpful in the field, and a good binocular microscope, with substage illumination, is essential for identification of worms, molluscs and crustaceans.

Begin identification with the quick-check key (fig. 39) which will enable you to identify the animal group to which your specimen belongs. The following few paragraphs then give brief descriptions of the major invertebrate animal groups of sandy beaches, and will direct you to the appropriate key. Each key is preceded by a brief reference to fuller literature sources. It must be stressed that these keys are designed for the identification of the resident intertidal sand macrofauna only. With a few exceptions, strictly sublittoral species have not been included, and dead shells and carapaces washed ashore from other habitats, and beached pelagic animals such as jellyfish, will not key out here. *The Collins Pocket Guide to the Seashore* (Barrett and Yonge, 1958) describes and illustrates all the common animals of the British seashore, including rocky shores and estuaries, as well as sandy shores, while *The Marine Fauna of the British Isles and North-West Europe* (Hayward and Ryland, 1990) covers some 2500 species, from the strandline to about 30 m depth. A companion volume to this one, *Animals on Seaweed* (Naturalists' Handbooks no. 9, Hayward, 1988) deals with the rich fauna of rocky shore seaweeds.

Notes on major animal groups occurring in sand

A. Sea anemones (Cnidaria: Zoantharia)

Most sea anemones live attached to a firm substratum and are primarily animals of the rocky shore. However, a few species have become adapted to burrowing in soft substrata, and some of these occur infrequently at ELWS on British coasts. They are generally deep burrowers, and are not easily collected. Removed from a sieve sample, they may not be immediately recognised as anemones, but will relax and eventually display their tentacles if left in a sand-floored aquarium. In shallow water on the lower shore their expanded tentacles can sometimes be spotted by a sharp eye. KEY I.

B. Ribbon worms (Nemerteans)

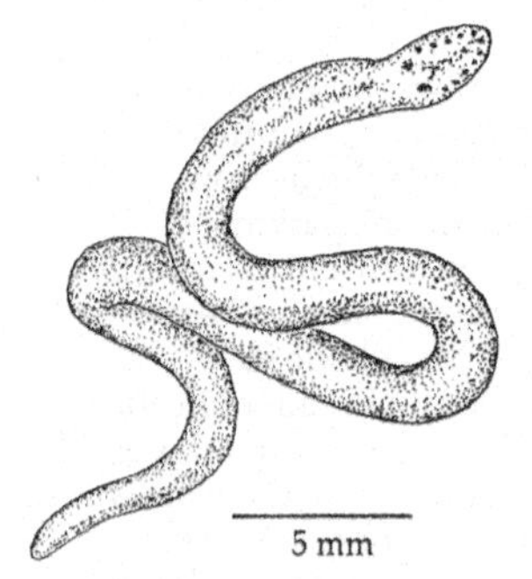

Fig. 40. A nemertean worm, *Lineus ruber* (Müller).

Flat, unsegmented worms without tentacles or any other projections. Sometimes brightly coloured, and often with groups of small, dark eyes visible on the head. In Britain the most spectacular species occur on rocky shores; some species occur not uncommonly in fine, muddy sands (fig. 40), but are often damaged during sieving and break up readily. The study of nemerteans requires special techniques, and the British species are still poorly known. Gibson (1982) describes all presently recognised British species, and provides details of techniques and procedures for their identification.

C. Bristle worms (Annelida: Polychaeta)

Polychaetes are generally abundant in all soft substrata, except for the coarsest sands. They are segmented worms, with each body segment bearing paired, biramous (two-branched) appendages, termed parapods, which may be variously modified according to the worm's mode of living. The two lobes of each parapod typically bear bundles or rows of bristles, or chaetae, which may be important in identification. The head frequently bears antennae, palps or tentacles, in various combinations, and there may be a conspicuous, eversible proboscis armed with powerful jaws. The sand fauna includes sedentary tube-dwellers, burrowers and fast-moving carnivores. KEY II.

D. Sipunculans

a sipunculan

Unsegmented, worm-like animals, with no chaetae, tentacles or other projections. There is no head, but a long, proboscis-like structure (the introvert) is protruded from the front end of the animal when feeding. The mouth lies at the tip of the introvert and may be surrounded by short tentacles. Sipunculans range in size from 1–40 cm. A few species occur rarely at ELWS on fine sandy shores. All British species may be identified using Gibbs (1977).

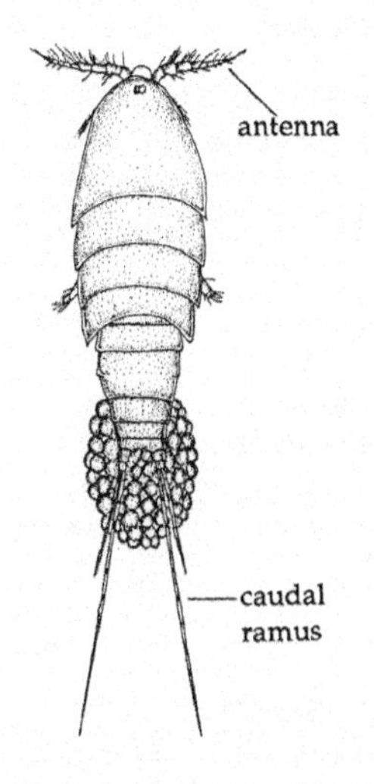

Fig. 41 *Harpacticus uniremis* Kröyer. Female with egg mass. Length 1 mm.

E. Harpacticoid copepods

Tiny crustaceans rarely longer than 1 mm. The body is slender, broadest at the front and tapered behind (fig. 41), with all body segments very similar. The first pair of antennae are very short and inconspicuous, and the second are longer and two-branched. Harpacticoids are abundant in soft substrata, including sand, and constitute one of the most important components of the meiofauna. Isaac (in Hayward and Ryland, 1990) provides keys to representative genera, but there is at present no modern handbook to the group.

F. Cumaceans

These distinctive little crustaceans might usefully be called "tadpole shrimps". The body, which rarely reaches 10 mm in length, consists of a swollen front portion – the carapace and pereon – and a slender pleon behind, terminated by a forked "tail" formed from the paired uropods (fig. 42). Most species live in the surface layers of sands and muds, from the intertidal zone into the deep sea, and a few species are common on British shores. KEY III.

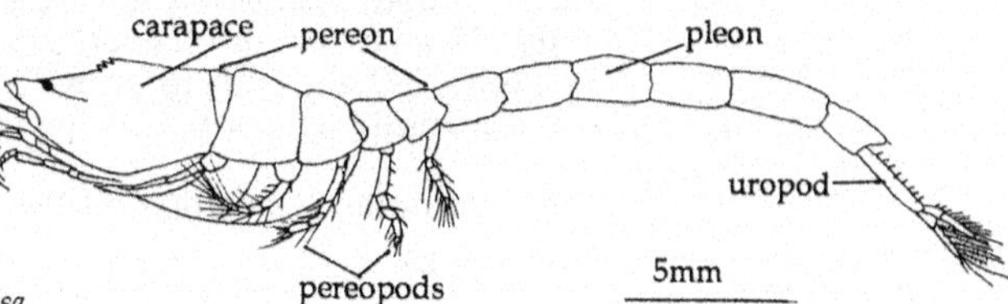

Fig. 42. A cumacean, *Iphinoe trispinosa*

G. Mysids

The "Opossum shrimps"or "Ghost shrimps" are abundant in coastal waters around Britain. On sandy shores they form part of the migrant night plankton and immense numbers may be netted on calm nights in the spring and summer. They are recognised by their slender, translucent bodies, prominent black eyes and feathery, two-branched limbs. KEY IV.

H. Isopods

Relatives of the Common Sea Slater (*Ligia*) can be common on some sandy beaches. The body is flattened when viewed from above, and usually rather elongate, with two pairs of antennae (the second pair longer than the first), and five to seven pairs of more or less identical thoracic limbs. KEY V.

I. Amphipods (Sandhoppers and their allies)

Small crustaceans, flattened from side to side, with two pairs of antennae, seven pairs of thoracic limbs (the first two pairs modified as gnathopods), three pairs of abdominal limbs and three pairs of uropods (VI.1). Amphipods are common, and often abundant on all sandy shores. KEY VI.

J. Shrimps and prawns (Decapoda: Natantia)

The "natant", or swimming, decapods include the Common Brown Shrimp. This (*Crangon crangon*) is familar to most seashore naturalists, but is sometimes confused with several less well-known species which may occur commonly on some beaches. KEY VII.

K. Crabs (Decapoda: Reptantia)

The "reptant", or walking, decapods include the ubiquitous Green Shore Crab, *Carcinus maenas* (pl. 5.3), which will be found on most sandy shores although it cannot be considered to be especially adapted to sandy shore living. Several species of swimming crab occur commonly in the surf community, and a number of other species are among the more obvious members of the macroinfauna. KEY VII.

L. Lamellibranchs or bivalves (Mussels, Cockles and their allies)

Both shallow-burrowing and deep-burrowing species may be abundant on all but the most exposed beaches. Tiny species, 2–3 mm long, recovered from sieves are often commensals of other burrowing infaunal species. KEY VIII.

M. Prosobranch gastropods (Snails,Whelks and their allies)

The few species found on sand include some ecologically important carnivores, and several spectacular pelagic species which may be cast ashore following south-westerly gales. KEY IX.

N. Opisthobranchs or Sea Slugs

Sand dwelling species have rather formless – truly slug-like – bodies with a reduced or internal shell. The animal crawls on a muscular foot; the upper surface is smooth or lobed, sometimes with conspicuous parapodial flaps, and the head may bear inrolled or pointed projections – rhinophores – at the front. KEY X.

O. Echinoderms (Starfish, Brittle stars and Sea urchins)

A few species of each of the major echinoderm groups are especially associated with sandy shores. KEY XI.

P. Hemichordates (Acorn Worms)

These elongate, flaccid, worm-like animals occupy deep burrows in silty sand at ELWS. None are common, and all are collected only by deep and careful digging. The body is unsegmented, but divided into distinct sections, with the front end loosely resembling an acorn within its cup. *Saccoglossus ruber* may exceed 200 mm long, with a very elongate front portion (pl. 7.5); *Glossobalanus sarniensis* (pl. 7.6) is longer, 500 mm or more, with a very short, acorn-like, front portion. Both occur rarely on southern and western coasts only.

Key I. Sea anemones

All British species of sea anemone are described by Manuel (1988), who has also published a colour guide (Manuel, 1983) illustrated by first class photographs. The few burrowing species found on British coasts require a sharp eye to detect them in the field. Four relatively common species are keyed out here.

1 Disc (flat, front end of animal) with about 70 tentacles, the outer, marginal series longer than the inner, typically banded with light brown. Body up to 150 mm long, tapered behind and with a pore at the end. Inhabiting a felt-like tube of mucus-bound silt. Sublittoral, but occasionally at ELWS. Common on all British coasts
Cerianthus lloydii Gosse (pl. 8.2)

– Disc with fewer than 30 short, unbanded tentacles. Animal not inhabiting a tube 2

2 With 12 tentacles only 3

– With 16 tentacles, each much longer than disc diameter. Body up to 70 mm long, pinkish to brown, with eight longitudinal rows of large, prominent warts. In mud and muddy sand, at LWST and shallow sublittoral. South and west coasts only
Edwardsia claparedii (Panceri) (pl. 8.5)

3 Tentacles longer than the diameter of the disc. Body up to 60 mm long contracted, 300 mm expanded; pinkish-brown, with brown spots. In sands and gravels, mostly sublittoral but occasionally at ELWS. All British coasts
Peachia cylindrica (Reid) (pl. 8.4)

– Tentacle length not exceeding disc diameter. Body up to 50 mm long (rarely larger), white, pink or pale brown with translucent longitudinal stripes. In sands and gravels; mostly sublittoral, sometimes at ELWS. All British coasts *Halcampa chrysanthellum* (Peach) (pl. 8.3)

Key II. Polychaetes

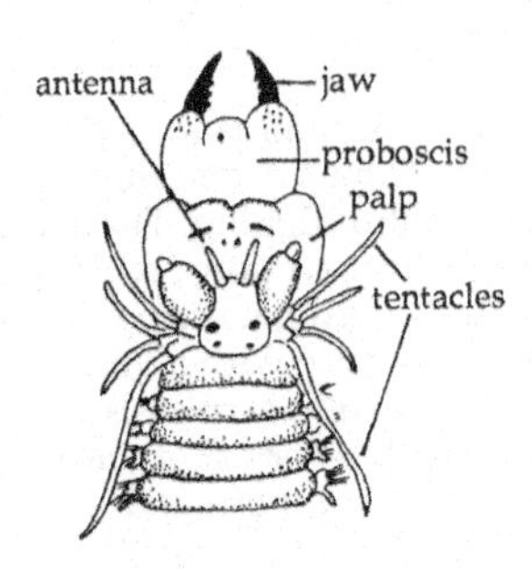

II. 1 *Perinereis cultrifera*, head and front end of body with proboscis everted

Polychaetes display a remarkable variety of body form, reflecting the equally wide variety of life modes adopted by numerous species present on British shores. In some the body segments are all essentially alike; in others the body is divided into distinct regions, consisting of two or more sequences of distinctly different types of body segment. In tube dwellers the front end usually comprises a fan of tentacles (pl. 3); the mouth may be flanked by prominent lips, but eyes and sensory appendages are lacking. Free-living species generally possess a well defined head, with eyes, antennae, palps, an eversible proboscis, and a variety of other appendages (II.1). Each body segment bears a pair of two-lobed appendages or parapods (II.2), which may be modified in various ways, and it is necessary to recognise the constituent parts of these appendages in order to identify many species. There is no modern monograph on the British polychaetes, but a well-illustrated account of the majority has been produced by Nelson-Smith and Knight-Jones (in Hayward and Ryland, 1990), and a number of families are now covered by *Linnean Society Synopses of the British Fauna*, numbers 32 (George and Hartmann-Schröder, 1985), 44 (Westheide, 1990) and 45 (Pleijel and Dales, 1991).

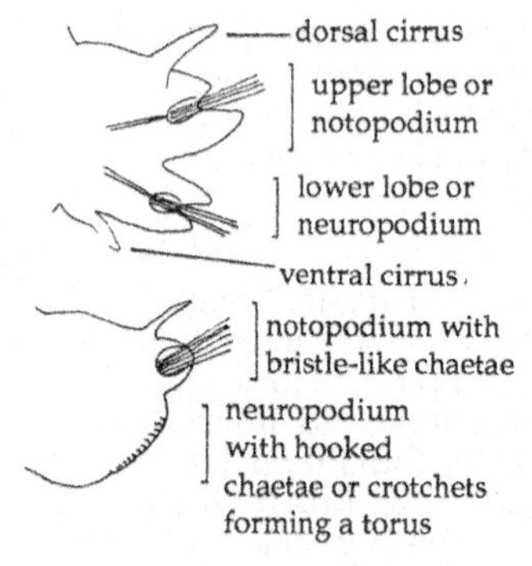

II. 2 Types of parapodia

1 Large oval animal, 10–20 cm long, the underside clearly segmented, indicating its annelid nature (II.3), the upper surface covered by a fine grey felt. Sides bristly, iridescent green and blue, with golden-yellow chaetae (bristles) visible. Sea Mouse: occurs offshore on muddy sand, often stranded on beaches, sometimes in large numbers *Aphrodita aculeata* (L.) (pl. 7.3)
– Not as described, more obviously worm-like 2

2 Upper surface of worm covered by flat, rounded scales, with fringed edges (II.4) [Scales may be lost as the animal is collected, leaving prominent mushroom-like stalks] 3
– Upper surface lacking scales 4

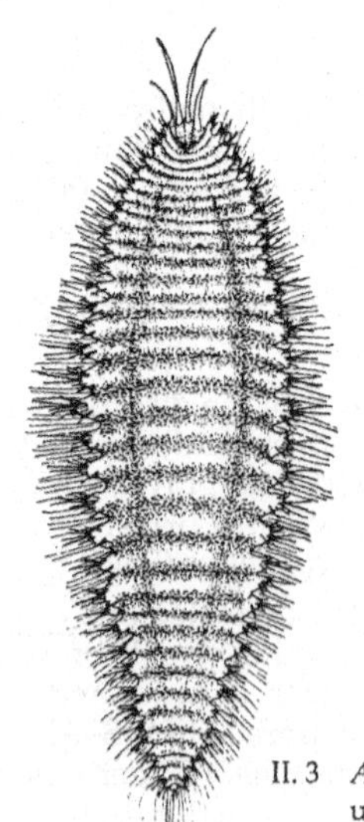
II. 3 *Aphrodita aculeata*, underside

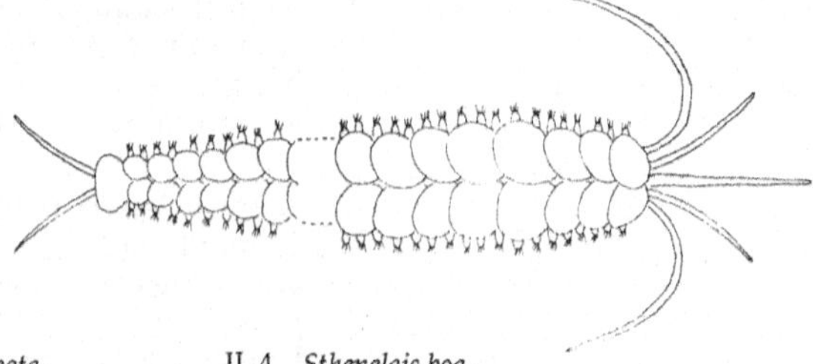
II. 4 *Sthenelais boa*, head and tail ends

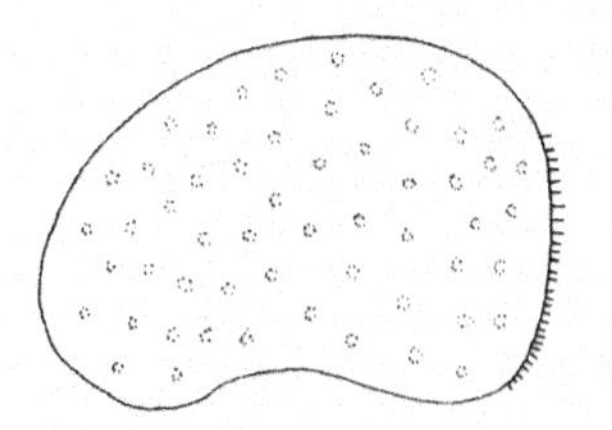
II. 5 Scale of *Sthenelais boa*

3 Scales covered with small warts and with a fringe of short, fine filaments (II.5). Head with a slender, median antenna (II.6). Up to 20 cm long, light grey to yellow. In muddy sand, on sheltered, mixed shores with stones and eel grass *Sthenelais boa* (Johnston) (II.6)

– Scales smooth and colourless, fringed with long, feathery filaments (II.7). No median antenna on head. Up to 15 cm long, greyish-white, iridescent underneath, with conspicuous longitudinal blood vessel *Sigalion mathildae* Audouin and Milne Edwards (II.7,8)

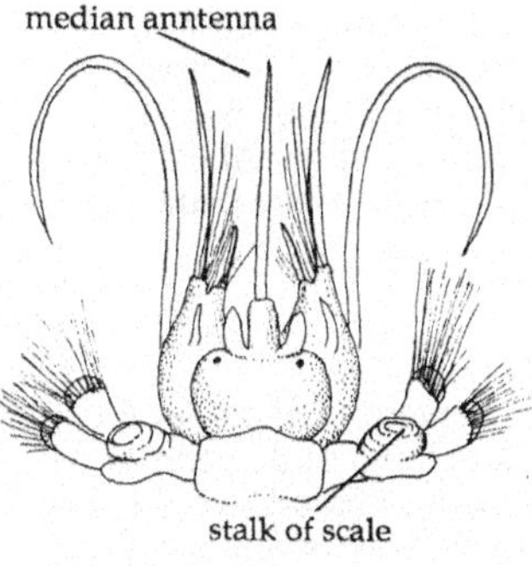

II. 6 *Sthenelais boa*, head and first body segment

4 Head of worm with long, well developed antennae, tentacles or palps (II.1, pl.3) 5

– Antennae, tentacles and palps short or absent 26

5 Head indistinct, the front end terminating in a crown of stiff, feathery tentacles (pl. 3.1), or a fan of stout bristles (pl. 3.4), or a coiling mass of slender, filamentous tentacles (pl. 7.1). Mostly tube-dwellers 6

– Front end of worm with variable number of antennae, palps and tentacles (II.1), but head clearly visible and usually well-developed. Mostly free-living 13

6 Body short and fat, divisible into three different regions. Front end blunt, with two fan-shaped clumps of thick golden-yellow spines. In a smooth, tusk-shaped tube of cemented sand grains, straight or slightly curved, 80 mm long. At LWS, on exposed beaches of fine to medium sand. All British coasts, locally abundant *Pectinaria koreni* Malmgren (pl. 3.4)

[*P. auricoma* (Müller) is found on more muddy shores, inhabiting a distinctly curved tube.]

– Not as described. Front end of worm with coiling, filamentous tentacles, or a feathery crown 7

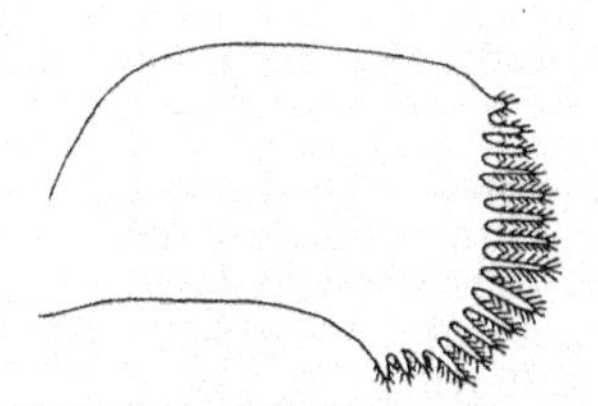
II. 7 Scale of *Sigalion mathildae*

7 Head of worm with a rigid, two-lobed fan of feathery tentacles 8

– Head of worm with a mass of coiling, filamentous tentacles 9

8 Worm up to 30 cm long, slender, living in a long, greyish tube of fine mud particles. Tentacle crown webbed at base, banded with red, brown and purple. On sheltered, mixed, muddy shores. At LWS, common on all coasts Peacock Worm *Sabella pavonina* Savigny (pl. 3.1)

– Worm up to 20 cm long, rather fat, in thick, gelatinous, partly transparent tube. In muddy sand, at LWS, south and west coasts only *Myxicola infundibulum* (Montagu) (pl. 3.5)

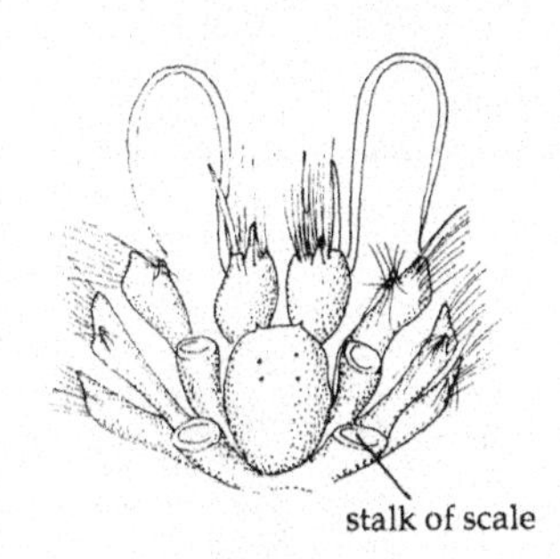

II. 8 *Sigalion mathildae*, head and first two body segments

II. 9 *Terebellides stroemi*, head end

9 A single gill present dorsally at the front end of the animal, consisting of a thick stem with four short branches, each with a row of fine plates (II.9). Up to 60 mm long, pinkish with bright red gills. In a membranous tube encrusted with sand or mud. ELWS and sublittoral, in muddy sand. All British coasts *Terebellides stroemi* Sars (pl. 7.1)

– Several pairs of gills dorsally at front end of animal 10

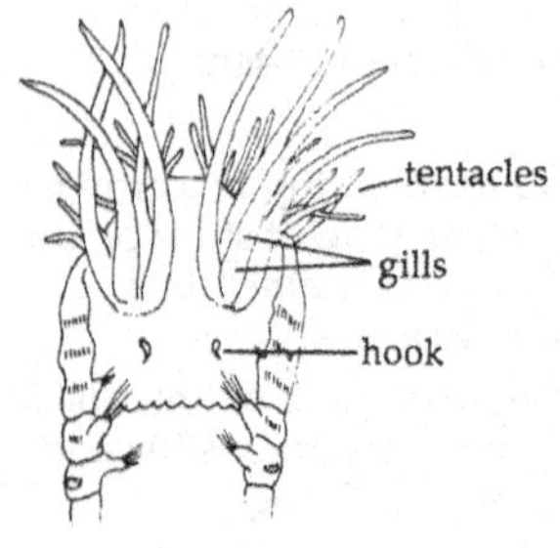

II. 10 *Melinna cristata*, dorsal view of head

10 Gills unbranched, arranged in two groups of four, with a conspicuous pair of hooks beneath. Tentacles on head thin and rather short. Up to 60 mm long, yellowish or pinkish-white, with gills flecked olive green. Lower shore and sublittoral, in membranous tube encrusted with mud particles, in mixed muddy sand. All British coasts *Melinna cristata* (Sars) (II.10)

– Gills branching, arranged in pairs 11

11 Two pairs of gills (II.11, 12) 12

– Three pairs of gills. Head with a pair of triangular lobes at base of tentacles. Up to 30 mm long, pink, yellowish or greenish, with white tentacles and blood red gills. In a stout tube of coarse sand grains, with a long fringe at the mouth which projects conspicuously above the sand surface. From MLWN into shallow sublittoral, on both exposed and sheltered shores, often abundant. All British coasts
Sand Mason *Lanice conchilega* (Pallas) (pl. 3.2)

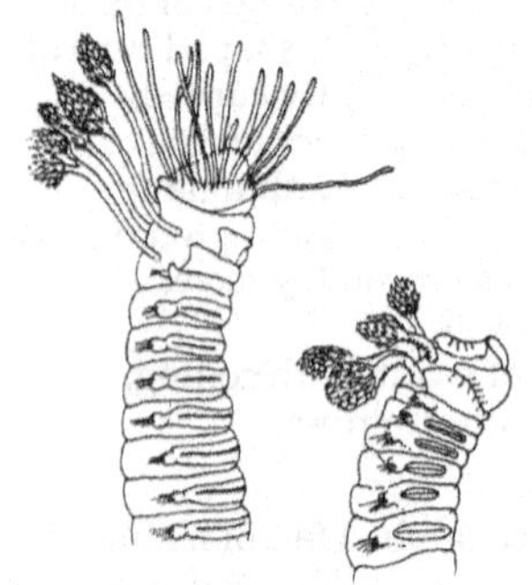

II. 11 *Pista cristata*, head end with tentacles extended (left) and contracted (right)

12 Gills with long, flexible stalks, branching at ends to give a bush-like form. Worm up to 90 mm long, dark red with brown gills. In membranous tube covered with shell fragments, sand grains and detritus. In fine muddy sand, lower shore and sublittoral. All British coasts *Pista cristata* (Müller) (II.11)

– Gills branching along most of their length (II.12). Worm up to 60 mm long, brick-red, speckled with white spots on upper surface. Gills bright red, tentacles violet. In delicate, transparent tube encrusted with sand grains and detritus *Nicolea venustula* (Montagu) (II.13)

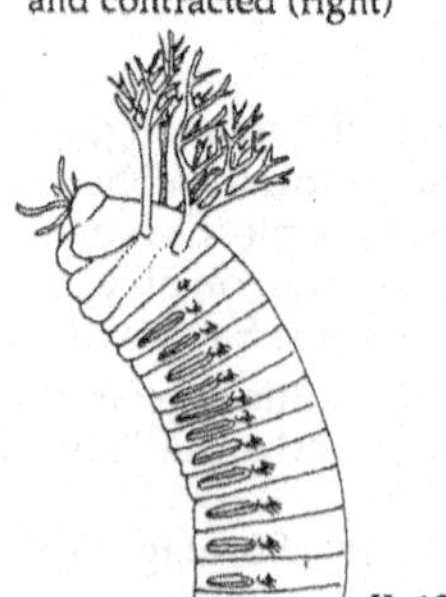

II. 12 *Nicolea venustula*, head end with tentacles contracted

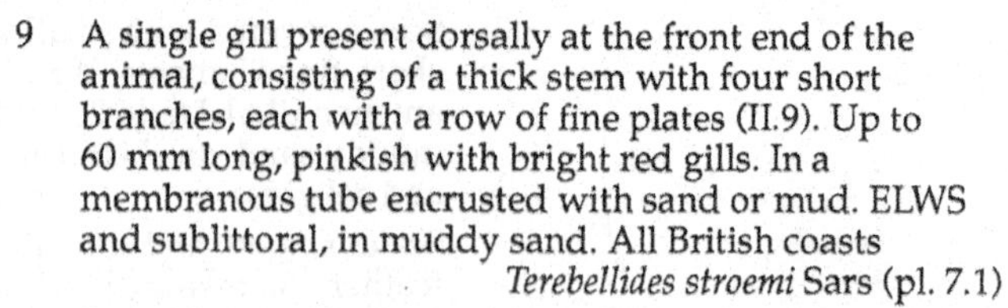

II. 13 *Nicolea venustula*, entire worm

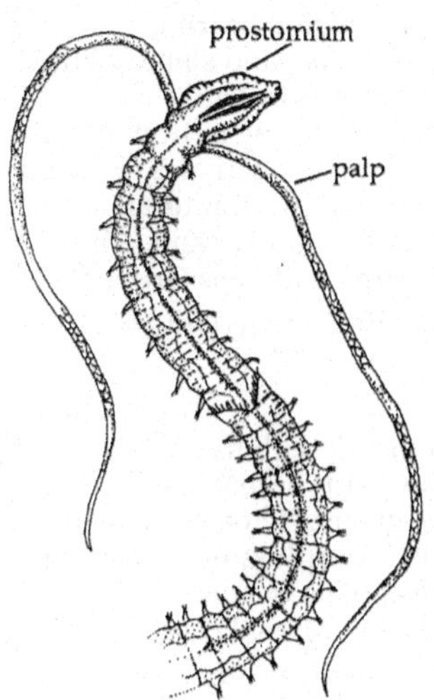

II. 14 *Magelona mirabilis*, head end of worm

13 Head with a single pair of long palps, very conspicuous, but easily broken off by careless handling; a few small, threadlike gills may be present, but other head appendages generally lacking (II.14–17) 14

– Head with variable number of appendages (sometimes numerous), not just a single pair 17

14 Head with a broad, flattened front portion (prostomium), resembling a duck's bill; palps ventrally attached, long and rather stout, fringed with fine papillae. No antennae or eyes. Body very thin: up to 170 mm long, only 2.5 mm wide; pink towards head end, greenish behind. At LWS and in shallow sublittoral, burrowing in clean sand. All British coasts
Magelona mirabilis (Johnston) (II.14)

– Head not as described; eyes usually present; palps rather small 15

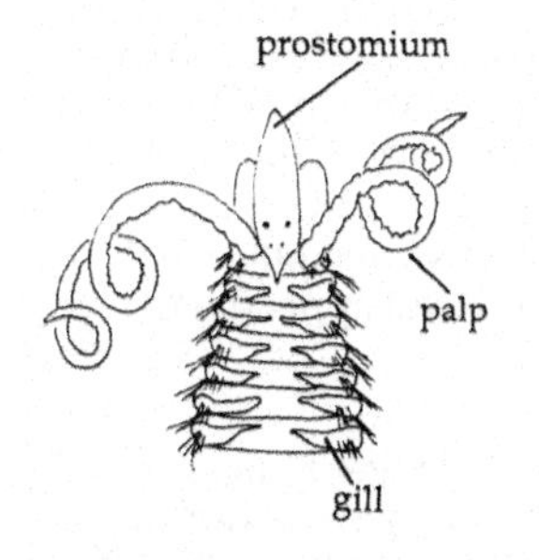

II. 15 *Scolelepis squamata*, head end

15 Elongate, fingerlike gills present on upper surface of worm, each usually with conspicuous blood vessel (II.15,16) 16

– Gills absent. Body thin: 60 mm long, 1.5 mm wide; pink at head end, darkening to red and green behind. Segments 5–15 have glandular bodies between the notopod and neuropod (II.2) on each side which secrete long, silky threads. At ELWS, in thin, sand-encrusted tube. All British coasts
Spiophanes bombyx (Claparède) (II.17)

16 Gills distributed along whole length of body. Prostomium ending in a blunt point behind the eyes, but without a median tentacle. Up to 8 cm long, 2–3 mm wide; bluish green, with bright red blood vessels; characteristically swims in spirals when disturbed. From MTL to LWS, on exposed beaches. All British coasts
Scolelepis squamata (Müller) (II.15)

– Gills missing from rear third of body. Prostomium with a short, conical, median tentacle behind the eyes. Up to 160 mm long, 6–9 mm wide, fragments readily. Red at head end, greyish green to transparent behind, with bright red gills. At LWS in fine, clean sand. All British coasts
Scolelepis foliosa (Audouin and Milne Edwards) (II.16)

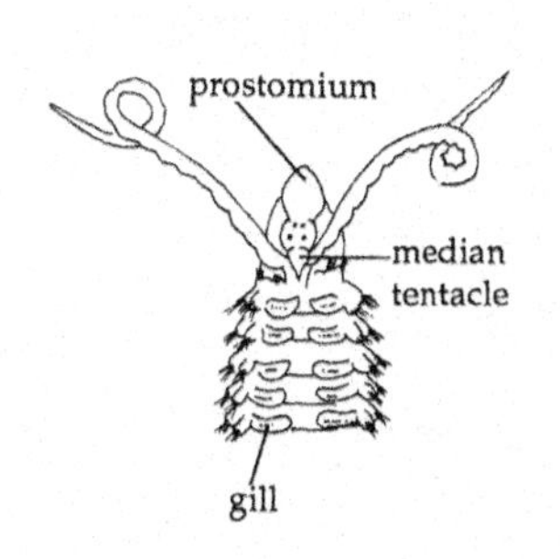

II. 16 *Scolelepis foliosa*, head end

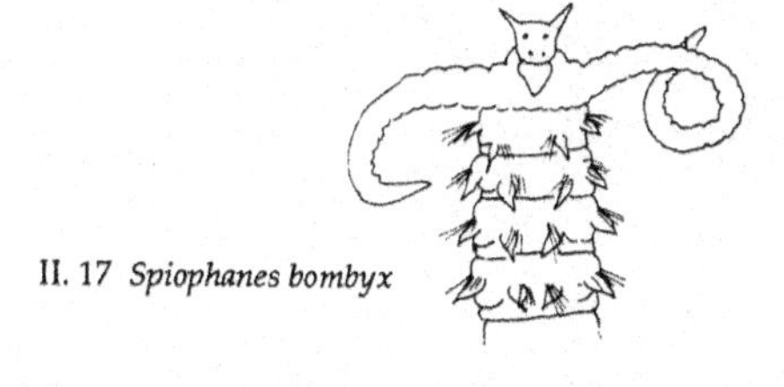

II. 17 *Spiophanes bombyx*

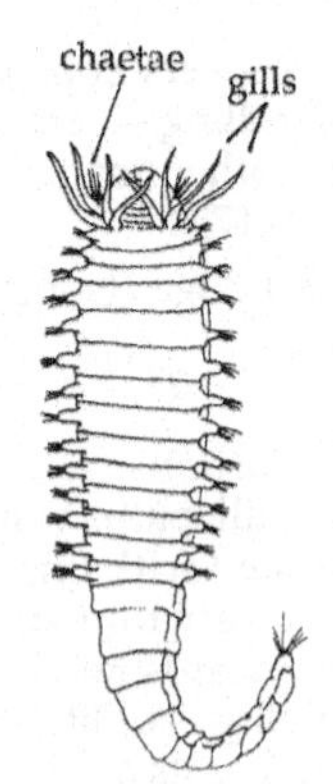

II. 18 *Ampharete acutifrons*

17 Body of worm with broad front region bearing prominent, two-branched parapodia, and slender hind region with rows of short hooks (uncini) only. A conspicuous bundle of golden chaetae in front of gills on each side of head. Up to 35 mm long, males greenish tinted, females pale pink. LWS and in sublittoral, in thin membranous tube among mixed, muddy sand. All British coasts *Ampharete acutifrons* Grube (II.18)

– Body of worm not distinctly differentiated into front and hind regions 18

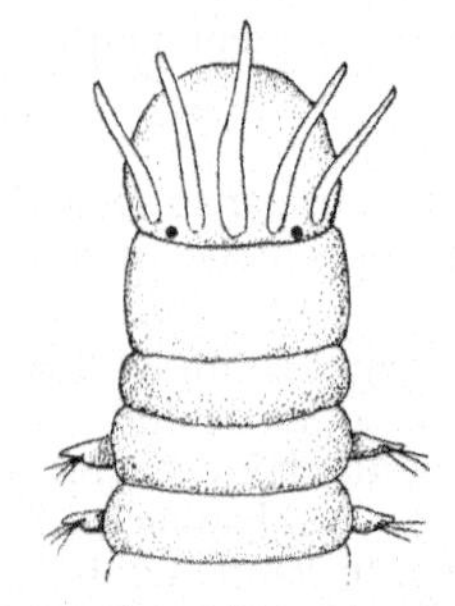
II. 19 *Marphysa belli*, head end

18 Head of worm smoothly rounded, with five short antennae, without prominent palps or tentacles. Comb-like gills present on segments 12 or 15 to about 35. Up to 200 mm long, pink or purple-grey, iridescent. At LWS, in muddy sand. Western coasts only *Marphysa belli* (Audouin and Milne Edwards) (II.19)

– Not as described; tentacles, palps and antennae variously developed, gills not comb-like 19

19 Parapods two-lobed, with dorsal and ventral lobes more or less equal in size (II.20,22). Head with two stout palps and four pairs of tentacles; proboscis, when everted, displays a pair of black jaws at end and small teeth on its surface (press the head gently from behind to evert the proboscis) (II.21,23) 20

– Parapods single-lobed. Proboscis with papillae and small teeth but not paired jaws. Head appendages various 22

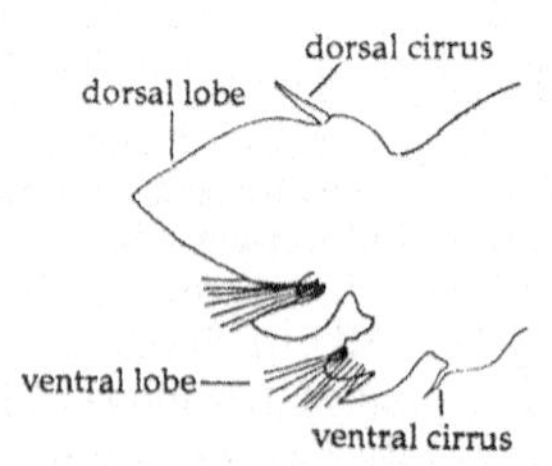

II. 20 Parapod of *Nereis virens*

20 Dorsal lobe of parapodium much longer than ventral, broadly leaf-like in shape, with a very small dorsal cirrus. A large, broad-bodied worm, up to or exceeding 300 mm, dark-green with a blue-green iridescence. Intertidal, in fine, muddy sand. All British coasts, but most common in south and west King Rag *Nereis virens* (Sars) (II.20,21)

– Dorsal and ventral lobes of parapodium similarly sized (II.22) 21

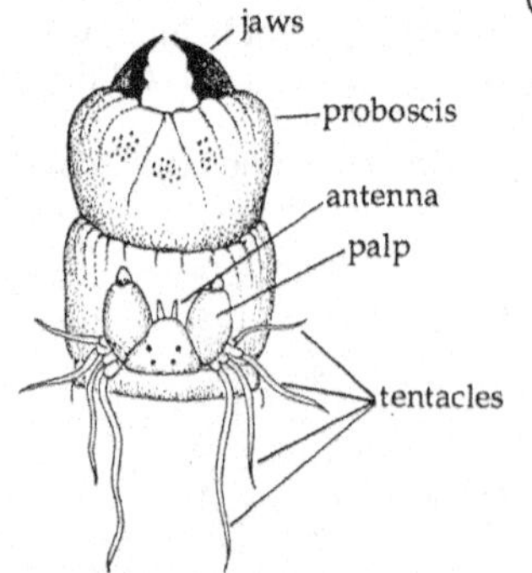

II.21 *Nereis virens*, head

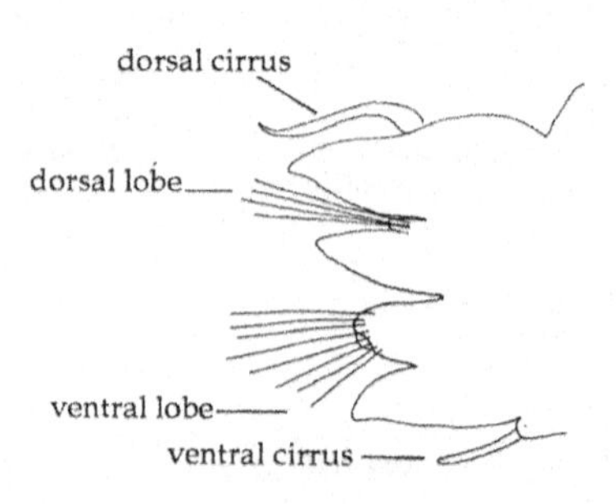

II. 22 Parapod of *Perinereis cultrifera*

PLATE 1

Some examples of the meiofauna

1. Tardigrade
   Batillipes phreaticus
   Renaud-Debyser
   (after Morgan
   & King, 1976)

2. Harpacticoid copepod
   Asellopsis hispida
   Brady
   (after Sars, 1911)

3. Ciliate
   Peritromus macronucleatus
   Ozaki & Yagui
   (after Wright, 1983)

4. Cnidarian
   Halammohydra vermiformis
   Swedmark & Teissier
   (after photographs
   in Clausen, 1967)
   (a) with posterior
   tentacle contracted
   (b) with posterior
   tentacle extended

5. Gastrotrich
   Turbanella
   (after Ruppert, 1988)

6. Sea slug
   Microhedyle lactea
   Hertling
   (after Thompson,1988)

7. Sea slug
   Hedylopsis suecica
   Odhner
   (after Thompson, 1988)

8. Polychaete
   Nerilla antennata
   O. Schmidt
   (after Westheide, 1990)

Scale lines:
50 μm for 1–5
1 mm for 6–8

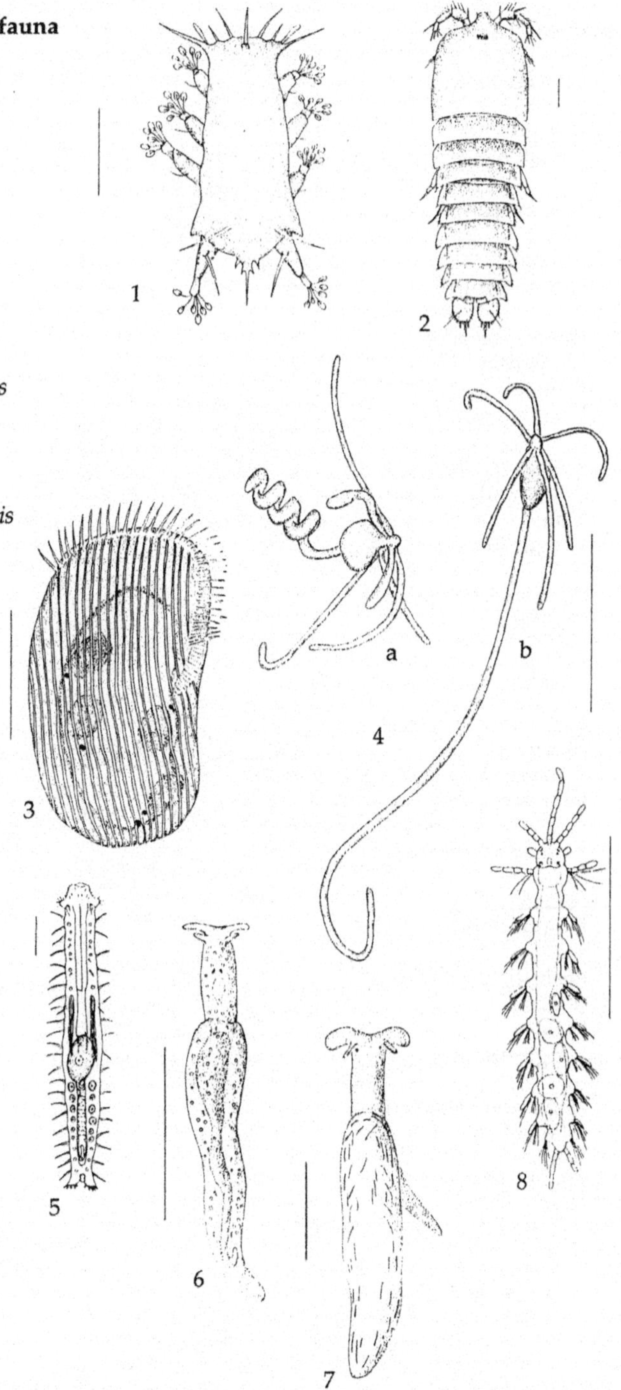

PLATE 2

Small crustaceans

1. *Eurydice pulchra* Leach
   Female.
   Length 6 mm
2. *Urothoe marina* (Bate)
   Female.
   Length 8 mm
3. *Bathyporeia pilosa* Lindström
   Female.
   Length 6 mm
4. *Talitrus saltator* (Montagu)
   Male.
   Length 20 mm
5. *Corophium volutator* (Pallas)
   Male.
   Length 8 mm
6. *Pontocrates arenarius* (Bate)
   Female.
   Length 6.5 mm
7. *Praunus inermis* (Rathke)
   Female.
   Length 15 mm
8. *Orchestia gammarellus* (Pallas)
   Male.
   Length 18 mm
9. *Haustorius arenarius* (Slabber)
   Female.
   Length 13 mm

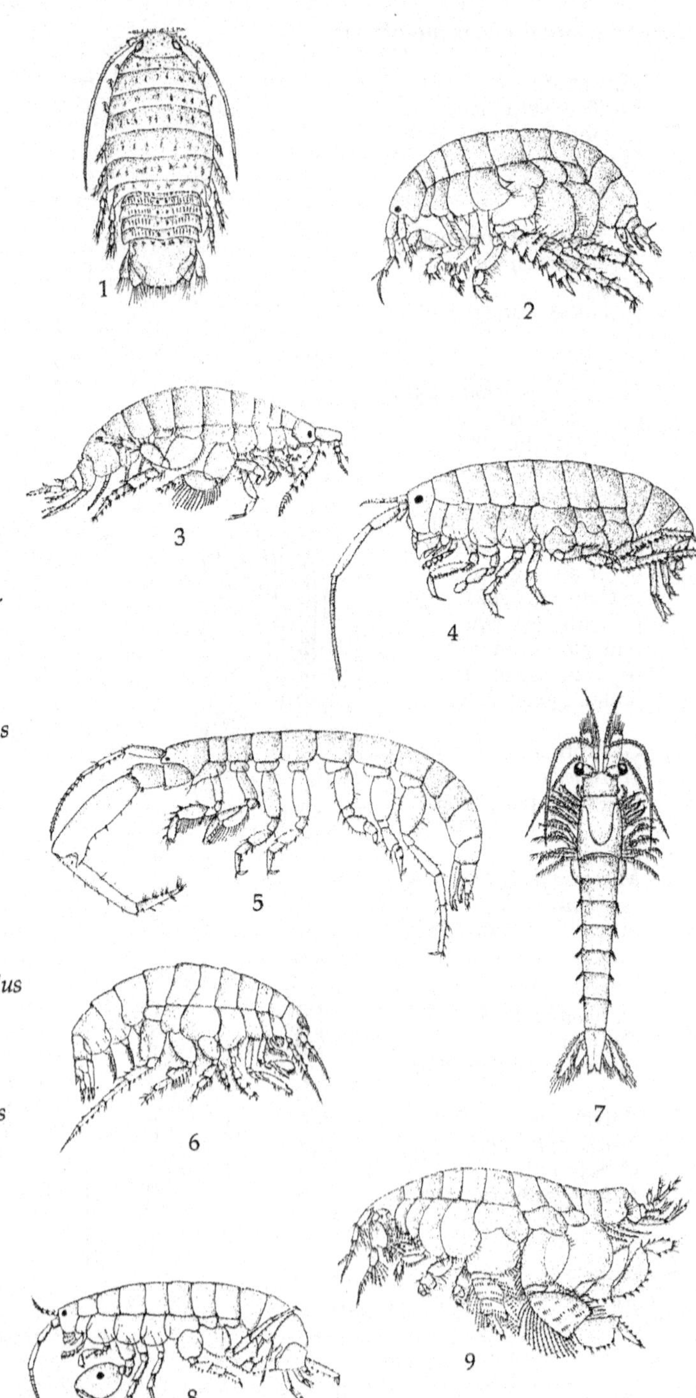

PLATE 3

Tube dwelling polychaetes

The head end of each worm is shown with the top end of the tube.

1. *Sabella pavonina* Savigny
2. *Lanice conchilega* (Pallas)
3. *Owenia fusiformis* delle Chiaje
4. *Pectinaria koreni* Malmgren
5. *Myxicola infundibulum* (Montagu)

Scale line: 1 cm

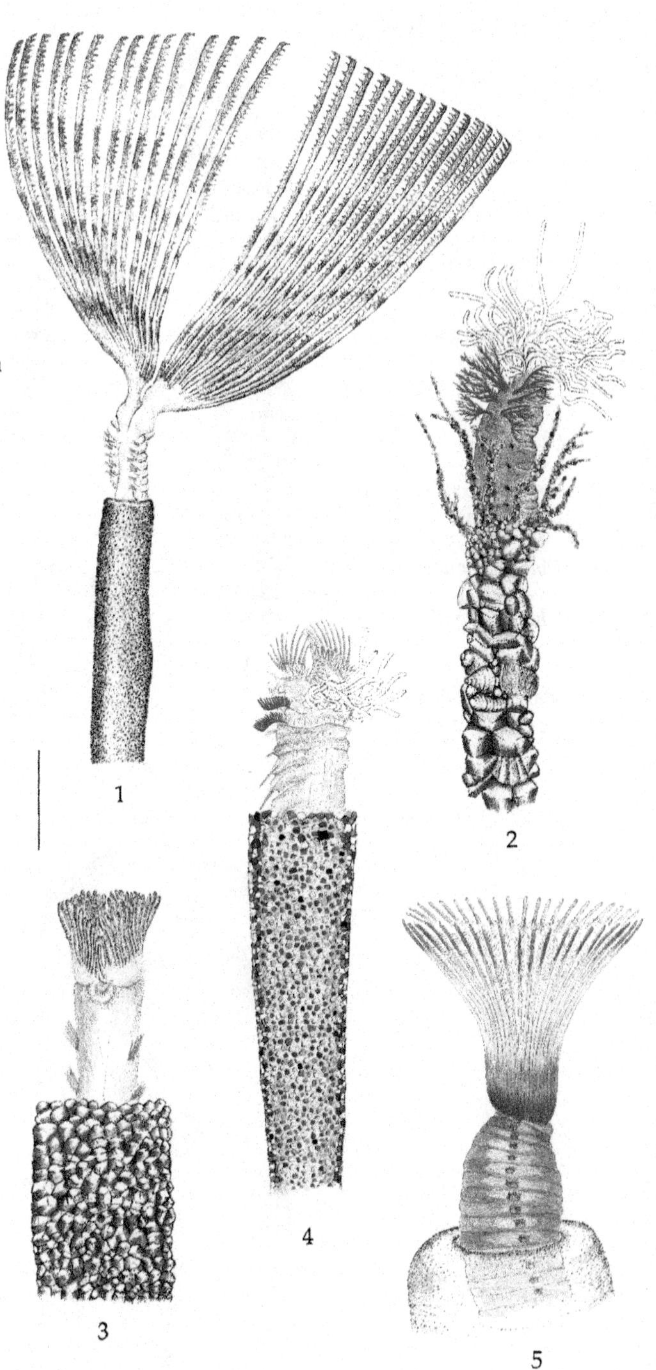

PLATE 4

Sandy shore molluscs

1. *Ensis arcuatus* (Jeffreys)
2. *Acteon tornatilis* (L.)
3. *Janthina janthina* L.
4. *Lunatia alderi* (Forbes)
5. *Lunatia catena* (da Costa)
6. *Donax vittatus* (da Costa)
7. *Angulus tenuis* (da Costa)
8. *Macoma balthica* (L.)
9. *Chamelea gallina* (L.)

All scale lines: 5 mm

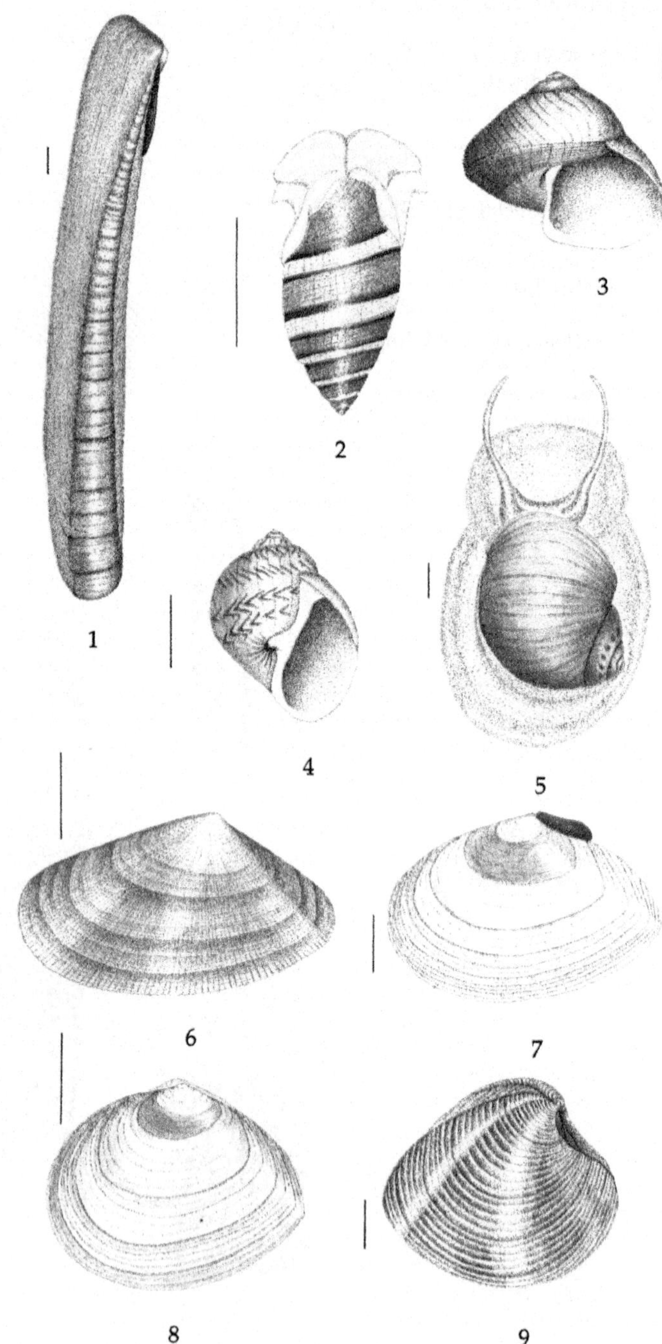

PLATE 5

Sandy shore crabs

1. *Corystes cassivelaunus* (Pennant) Male
2. *Portumnus latipes* (Pennant)
3. *Carcinus maenas* (L.)
4. *Thia scutellata* (Fabricius)
5. *Liocarcinus depurator* (L.)
6. *Liocarcinus marmoreus* (Leach)

All scale lines: 1 cm

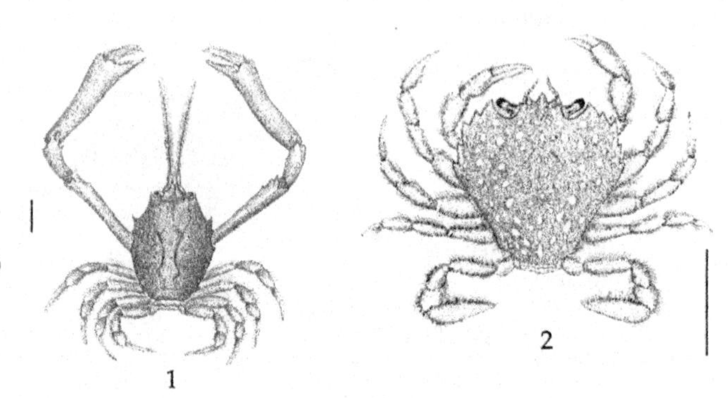

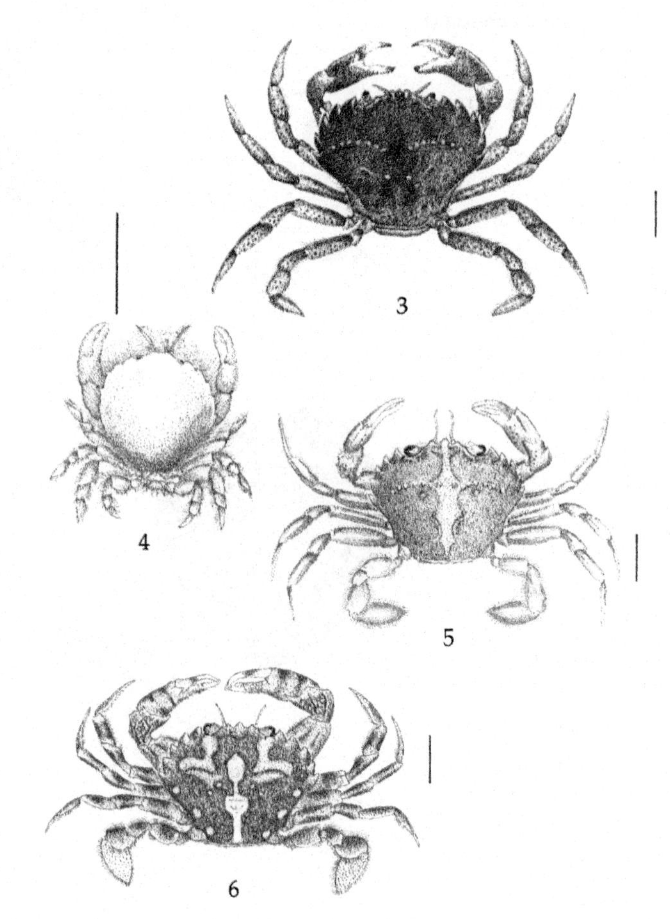

PLATE 6

Sandy shore echinoderms

1. *Asterias rubens* L.
2. *Echinocardium cordatum* (Pennant)
3. *Amphiura brachiata* (Montagu)
4. *Astropecten irregularis* (Pennant)
5. *Spatangus purpureus* O.F. Müller
6. *Ophiura ophiura* (L.)

All scale lines: 5 mm

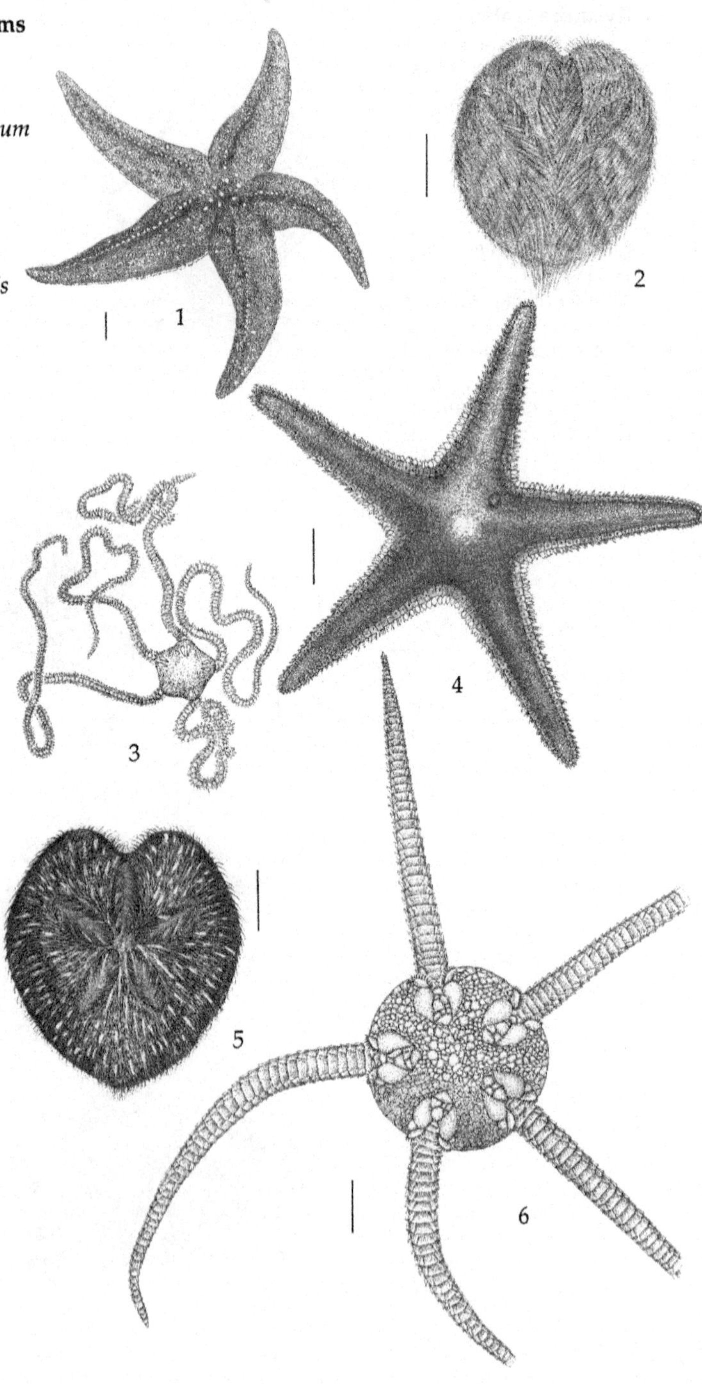

PLATE 7

Worms and worm-like animals

1. Polychaete
   Terebellides stroemi Sars

2. Polychaete
   Arenicola marina (L.)

3. Polychaete
   Aphrodita aculeata (L.)

4. Sea cucumber
   Leptosynapta inhaerens
   (O.F. Müller)

5. Hemichordate
   Saccoglossus ruber
   (Tattersall)

6. Hemichordate
   Glossobalanus sarniensis
   Koehler

All scale lines: 1 cm

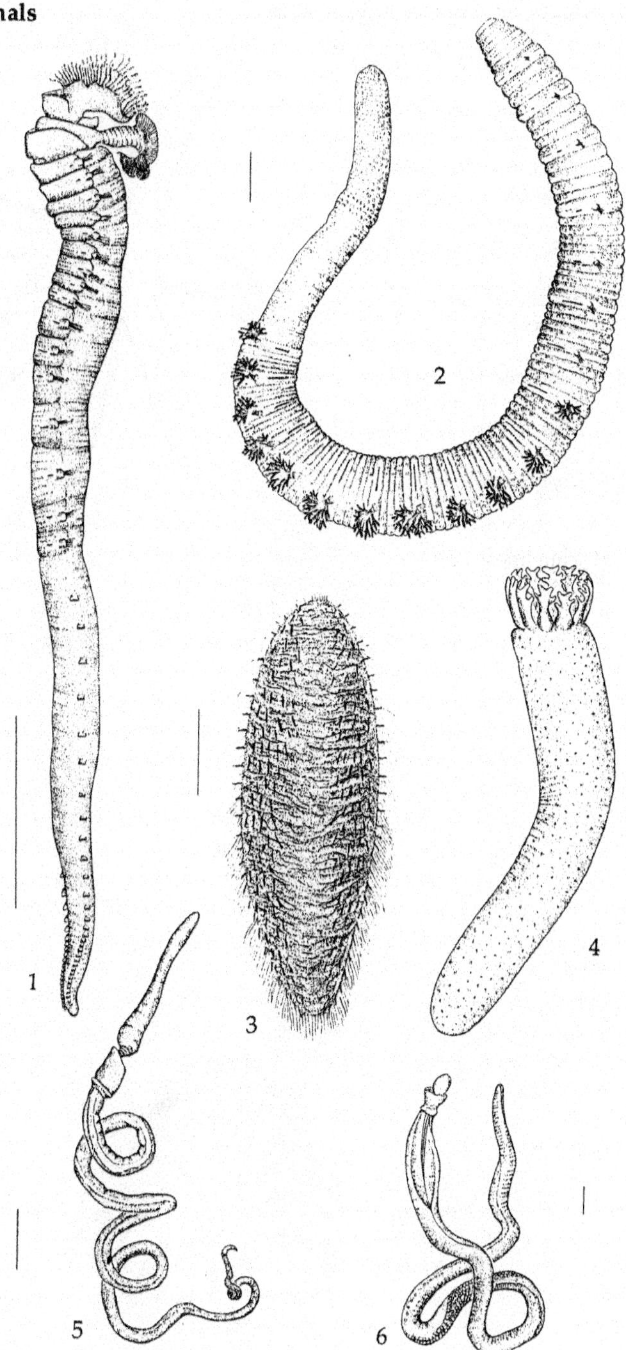

PLATE 8

Molluscs, cnidarians and a starfish

1. *Luidia ciliaris* (Philippi)
2. *Cerianthus lloydii* Gosse
3. *Halcampa chrysanthellum* (Peach)
4. *Peachia cylindrica* (Reid)
5. *Edwardsia claparedii* (Panceri)
6. *Cerastoderma edule* (L.)
7. *Mya arenaria* L.
8. *Mya truncata* L.

All scale lines: 1 cm

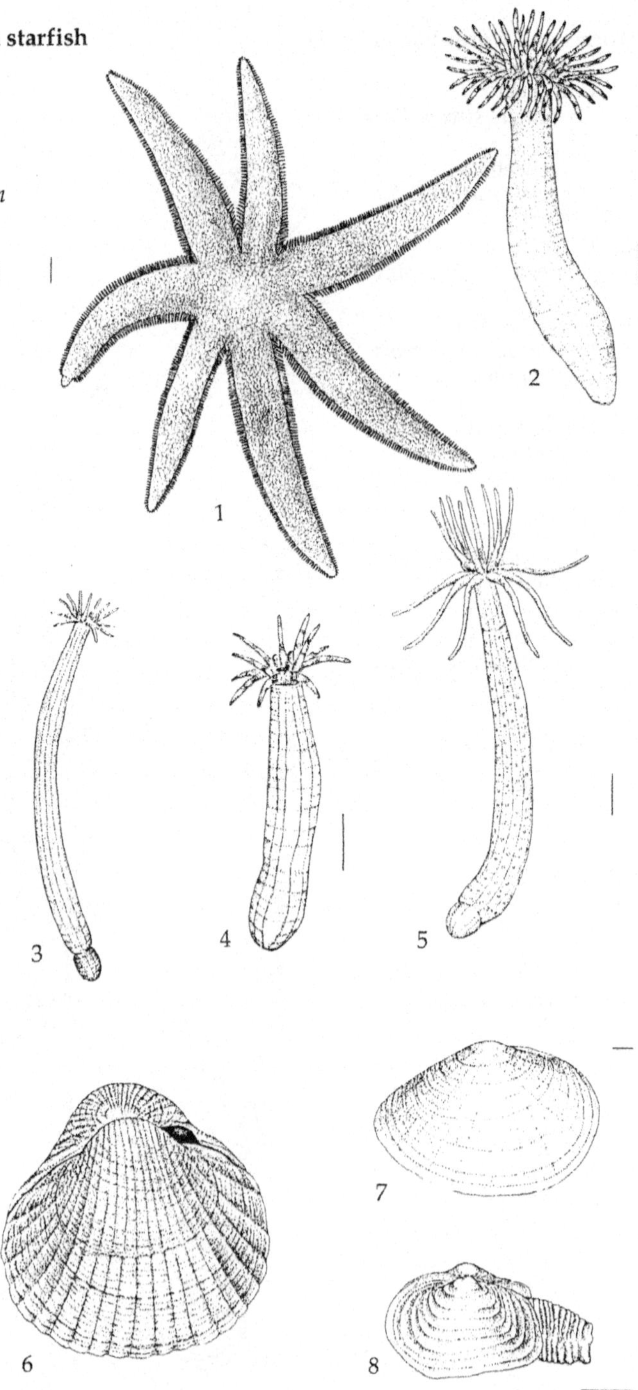

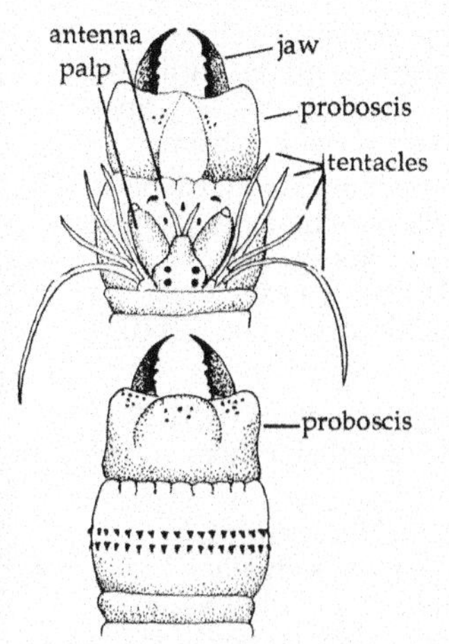

II. 23 *Perinereis cultrifera*, head viewed from above (top) and below (bottom)

21 Dorsal cirrus longer than dorsal lobe of parapodium (II.22). Head with two antennae, slightly shorter than palps; posterior pair of tentacles longer than others. Proboscis with two curved groups of tiny black teeth on upper side. Up to 250 mm long, greenish-bronze, with distinct longitudinal blood vessel. On mixed shores, under stones and among eelgrass. All British coasts
Rag Worm *Perinereis cultrifera* (Grube) (II.22,23)

– Dorsal cirrus shorter than dorsal lobe of parapodium (II.24). Head with two antennae, much shorter than palps, and tentacles about as long as body width. Up to 120 mm long, greenish to orange, with conspicuous longitudinal blood vessel. Intertidal, in fine muddy sand, tolerant of brackish water. All British coasts
Rag Worm *Nereis diversicolor* O.F. Müller (II.24,25)

22 Head with three small antennae, with a group of three tiny eyes on each side. Palps fused along their length to form a conspicuous conical structure. A small, thin worm, up to 10 mm long, creamy white. At LWS and in sublittoral, in shelly or gravelly sands. All British coasts
Exogone hebes (Webster and Benedict) (II.26)

– Head with four or five antennae, and two to four pairs of tentacles. Parapodia characteristically with large, paddle-shaped dorsal cirri 23

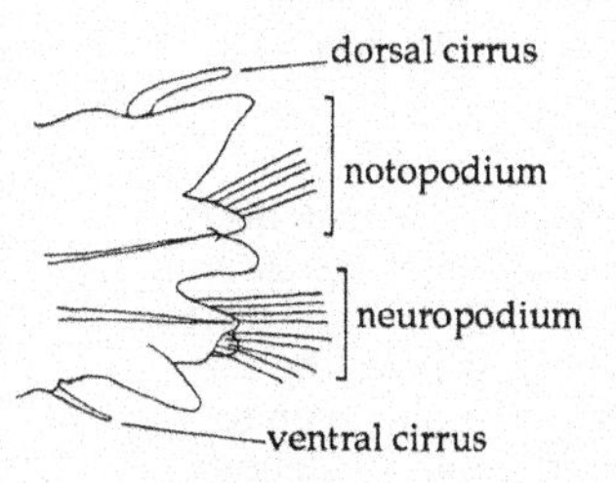

II. 24 *Nereis diversicolor*, parapod

23 Two segments immediately behind head lack paddle-shaped cirri; the first of these has instead two pairs of short, conical tentacles (II.27) 24

– Three segments immediately behind head without paddle-shaped cirri; instead, four pairs of tentacles (II.29,30) 25

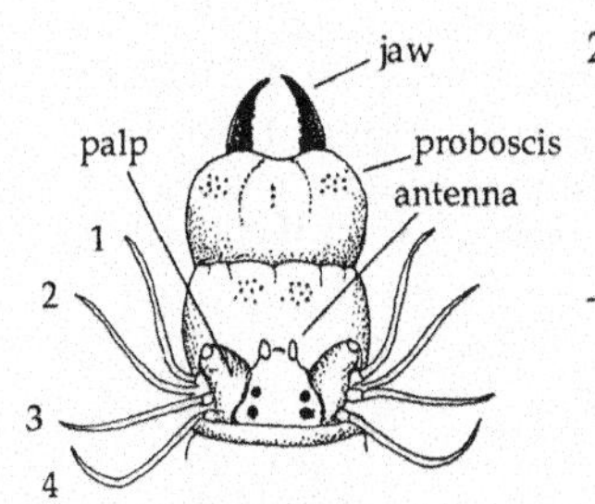

II. 25 *Nereis diversicolor*, head
1–4 = tentacles, paired

24 Parapodium of segment 2 with a ventral cirrus, but no chaetae. Up to 300 mm long, yellowish white with indistinct purplish spots along sides. In clean sand, at LWS. Southern and western coasts only
Eteone foliosa Quatrefages (Fig.II.27)

– Parapodium of segment 2 with chaetae as well as ventral cirrus. Up to 120 mm long, yellow to brick-red. At LWS and in shallow sublittoral, all British coasts
Eteone flava Fabricius (II.28)

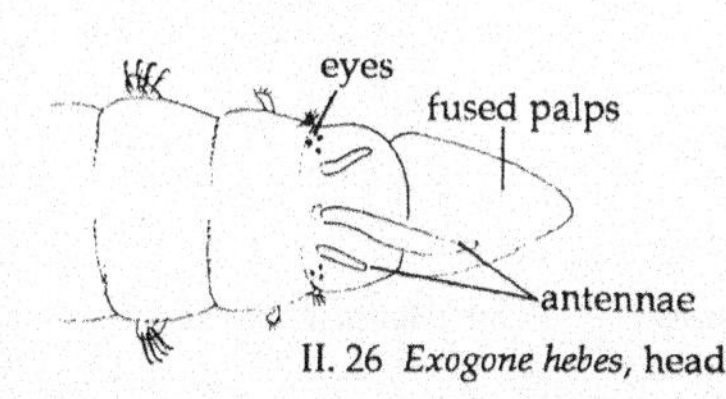

II. 26 *Exogone hebes*, head

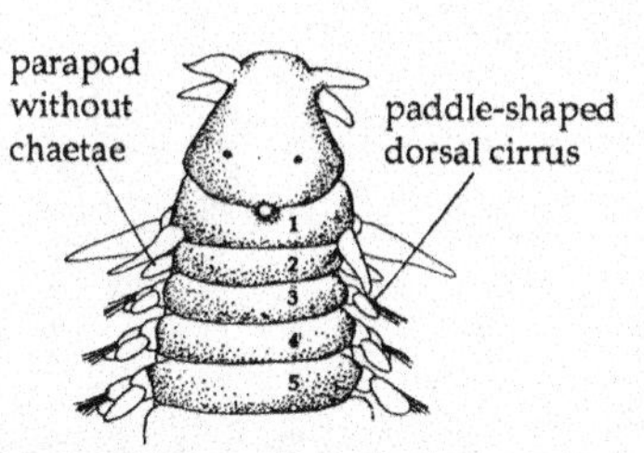

II. 27 *Eteone foliosa*, head and first five body segments

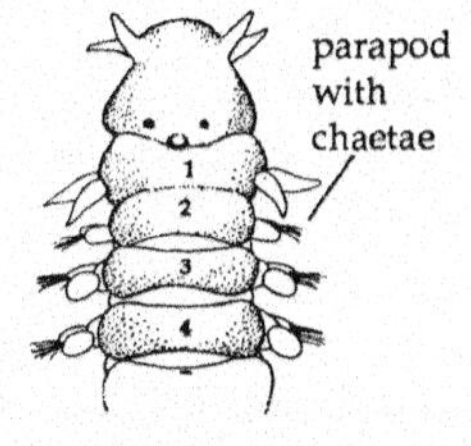

II. 28 *Eteone flava*, head and first four body segments

25 Large worm, up to 300 mm long. Greenish-yellow, banded with brown and blue; paddles green, with large brown blotches. LWS and shallow sublittoral, all British coasts *Phyllodoce groenlandica* Bergström (II.29)

– Smaller, usually less than 100 mm long. Yellowish or off-white, each segment with three brown blotches, giving a banded effect; paddles brown. In muddy sand, particularly under stones. All British coasts *Phyllodoce maculata* (L.) (II.30)

26 Body of worm divisible into distinct regions (II.31, 32), or with some segments much longer than others. Mostly tube-dwelling 27

– Body not obviously divided into distinct regions. Segments all roughly the same size, or gradually increasing in size towards middle of worm, but none obviously much longer than others. Usually free living 37

27 Some body segments very much longer than wide, the head end broadly rounded or obliquely flattened 28

– Most body segments shorter than wide, often with many fine rings. Head tapered or conical, like that of an earthworm 31

28 Longest segments towards head end of worm. Body tapered behind, the segments becoming progressively smaller towards the simple, rounded tail end. Head with six frilled membranous lobes surrounding the terminal mouth. Up to 100 mm long, pale green to yellowish, the head end tinted red; builds a slender mucus tube encrusted with shell fragments and sand grains. From MLWS into shallow subtidal, in fine to medium sand. Common on all British coasts *Owenia fusiformis* delle Chiaje (II.31, pl.3.3)

– Largest segments towards hind end of animal. Not tapered behind, the body ending in a funnel shape, or broadly rounded, or abruptly truncated (as if cut off) 29

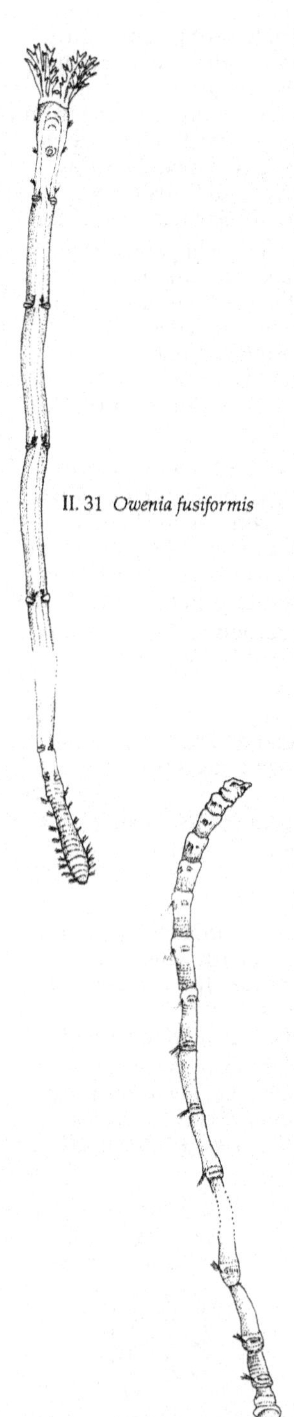

II. 31 *Owenia fusiformis*

II. 32 *Nicomache lumbricalis*

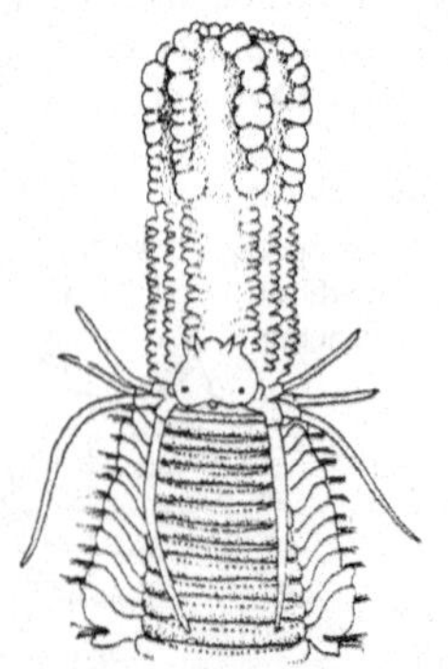

II. 29 *Phyllodoce groenlandica*, with proboscis everted

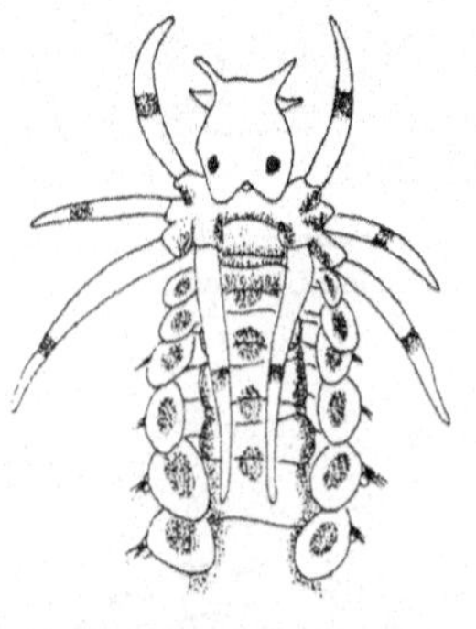

II. 30 *Phyllodoce maculata*

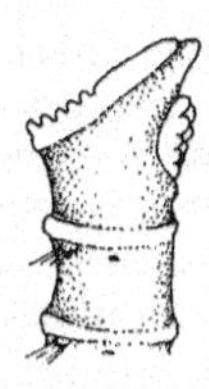
II. 33 *Euclymene lumbricoides*, head end

29 Head end of worm broadly rounded, hood-like; tail end a short funnel with a fringed rim of 15–25 short lobes. Up to 160 mm long, tinted reddish brown towards head; in a thick tube encrusted with sand grains. At LWS and in the shallow sublittoral, all British coasts
Nicomache lumbricalis (Fabricius) (II.32)

– Head end of worm obliquely truncated, in the form of a flat plate with a notched or toothed rim (II.33) 30

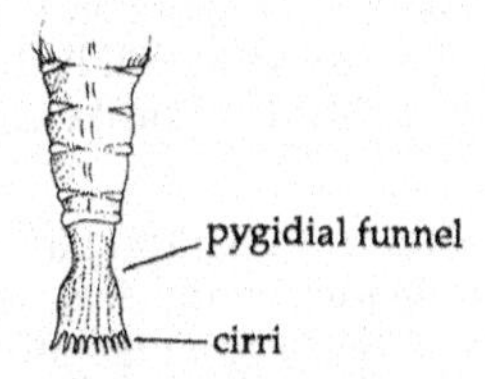

II. 34 *Euclymene lumbricoides*, tail

30 Hind end of worm a short pygidial funnel (II.34), fringed with cirri, with the terminal anus within. Up to 150 mm long, pale brown or pinkish, with red and white bands. In a thick tube encrusted with shell fragments and small stones. At LWS, usually on mixed, sheltered shores with eelgrass and stones. Western coasts only
Euclymene lumbricoides (Quatrefages) (II.33, 34)

[*E. robusta* Arwidsson is similar, but has an asymmetrical pygidial funnel, with a few cirri much longer than the rest (II.35). On western coasts only.]

– Hind end of worm a flattened plate with a notched rim (II.36). Anus not terminal, but dorsal, above the edge of the plate. Up to 100 mm long, dark brown towards head, lighter towards tail; in a mud-encrusted tube. At LWS and in shallow sublittoral, in mixed shelly or muddy sands. Northwest and North Sea coasts
Maldane sarsi Malmgren (II.36, 37)

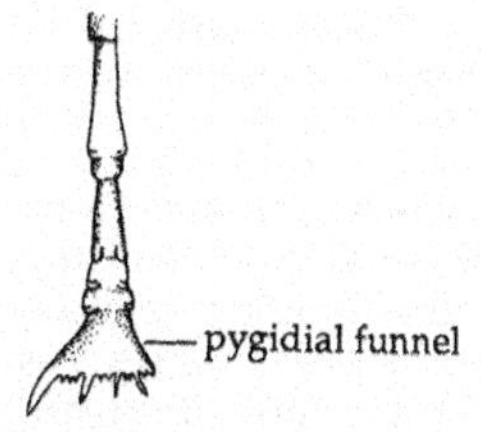

II. 35 *Euclymene robusta*, tail

31 Body with a thick front region of 19 ringed segments, the last 13 of which each bears a pair of branching red gills; hind region much thinner than front, without gills or chaetae, often missing or damaged. Up to 200 mm long; pink when small, then dark yellowish green, the largest specimens black. Lower shore, occupying permanent burrows in fine to medium sand; all coasts, often abundant Lugworm *Arenicola marina* (L.) (pl. 7.2)

[The so-called Black, or Laminarian, Lugworm which occupies the lowest part of the beach is now known to be a species distinct from the paler coloured Blow Lug, which ranges further up the shore. The Black Lug, *A. defodiens* (Cadman and Nelson-Smith (1993), lives in a deep J-shaped tube.]

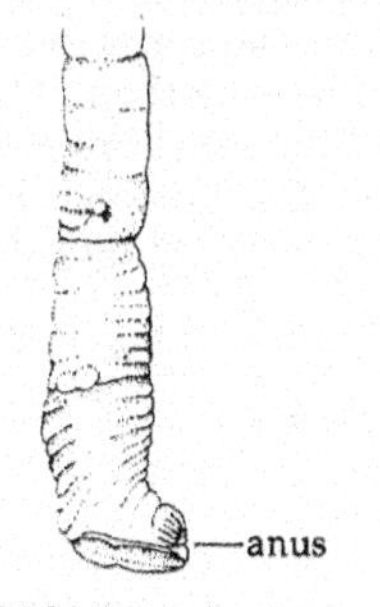

II. 36 *Maldane sarsi*, tail

– Front region of worm short and broad, with short segments, but lacking branched gills. Hind portion longer, with more elongate segments 32

32 Parapods often small, with fine chaetae only (II.2) 33

– Some parapods with dorsal and/or ventral hooked chaetae (crotchets) arranged in rows (tori) resembling zip fastener (II.2) 34

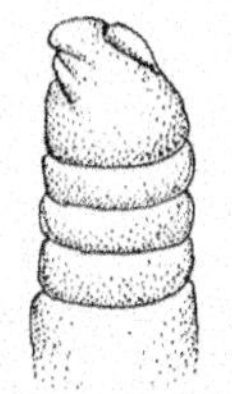
II. 37 *Maldane sarsi*, head

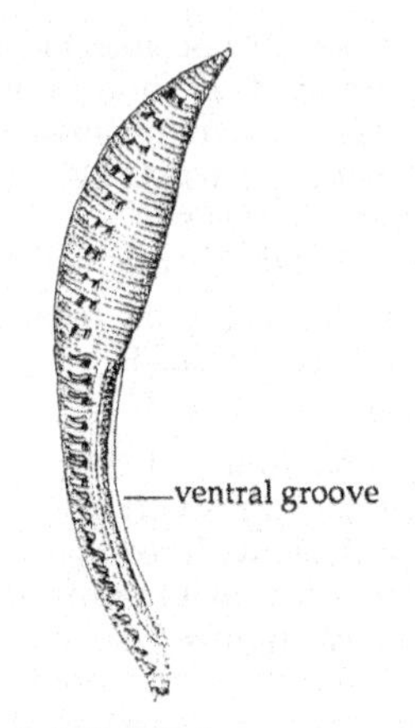

II. 38 *Ophelia rathkei*

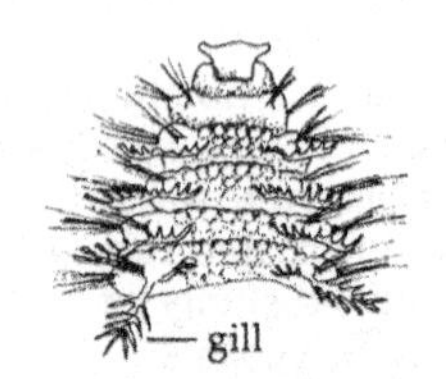

II. 39 *Scalibregma inflatum*

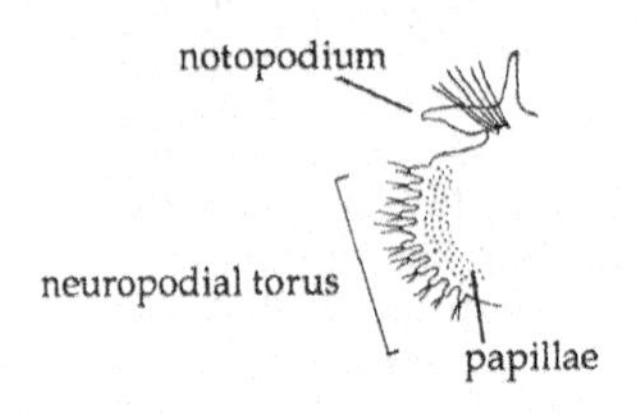

II. 40 *Orbinia latreilli*, parapod viewed from behind

33 Body comprising a broad front region, and a narrower hind region with a deep, ventral groove. 15 pairs of long, slender, paddle-shaped gills on hind segments. Up to 10 mm long; pink, with iridescent sheen. Lower shore and shallow sublittoral, in medium to coarse sand. Probably all coasts *Ophelia rathkei* McIntosh (II.38)

– Body broad in front, narrow behind, but without a ventral groove. Four pairs of branching gills on first four segments behind head; slender dorsal and ventral cirri on all but first 16 segments. Up to 100 mm long, purplish red with yellow patches. Lower shore and shallow sublittoral, in fine and muddy sand. All British coasts *Scalibregma inflatum* Rathke (II.39)

34 Front part of body broad, flattened, with parapodia along the sides; hind part more slender, cylindrical, with parapodia on the upper surface. Front segments with conspicuous rows of crotchets. Broad, simple gills present on upper surface of all segments except first few 35

– Front part of body broad and wrinkled, without crotchets. All parapodia generally indistinct 36

35 Gills present on upper surface of all segments from number 5 backwards. Rows of papillae form a fringe behind the neuropodial torus, on hindmost segments of front region (II.40). Up to 400 mm long; pinkish towards head, yellowish towards tail. Lower shore, in clean sand. South and west coasts only
Orbinia latreilli (Audouin and Milne Edwards) (II.41)

– Gills present on upper surface of all segments from numbers 9 to 17 backwards. Neuropodial torus in front region with up to three papillae, or lacking them. Up to 20 mm long, bright pinkish orange. Lower shore and shallow sublittoral, in fine muddy sand. North and west coasts *Scoloplos armiger* (Müller) (II.42)

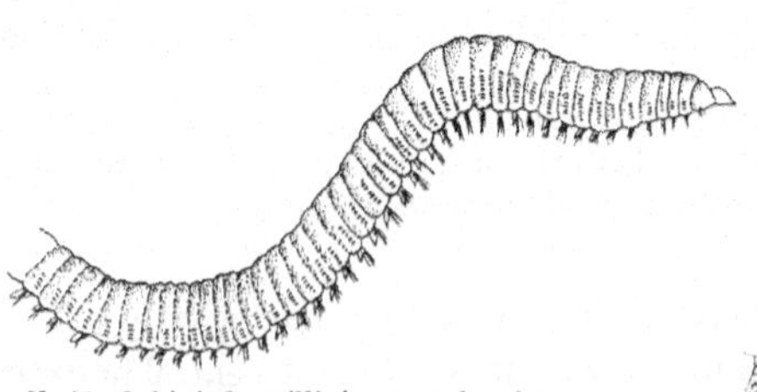

II. 41 *Orbinia latreilli*, front end only

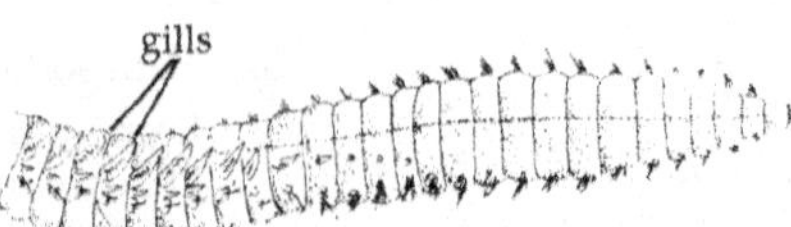

II. 42 *Scoloplos armiger*, front end only

36 Front part of body consisting of head plus nine or ten chaetae-bearing segments. Body long and thin, up to 100 mm, blood red in colour. LWS and shallow sublittoral, in mud or muddy sand, especially in polluted situations. Common on all coasts
Capitella capitata (Fabricius) (II.43)

– Front part of body consisting of head plus 12 chaetae-bearing segments. Up to 300 mm long, very thin and fragile; deep red to purple at head end, light yellowish towards tail. LWS and shallow sublittoral, in fine sand and muddy sand
Notomastus latericeus (M. Sars) (II.44)

37 Slender, cylindrical worms with numerous ringed segments. Head end with a long, tapering prostomium with four minute tentacles at its tip (II.45). When active, will rapidly evert and withdraw a conspicuous balloon-like proboscis, with conspicuous jaws at tip. Vigorous when newly collected 38

– Not like this 41

38 Mouth ringed with numerous jaws. A row of 7–11 chitinous chevrons towards base on each side of proboscis. Up to 100 mm long; pale green to yellow, orange towards tail, with brown flecks. At LWS and sublittoral. All British coasts
Goniada maculata (Oersted) (II.45)

– Mouth with just four large, black fangs (II.46). No chitinous chevrons on proboscis 39

39 Proboscis covered with small warts, many with a dark tip, resembling a fingernail 40

– Proboscis covered with rounded or conical warts, none of which have a chitinous tip. Prostomium with 10–12 rings. Up to 200 mm long, pinkish red. Lower shore, in fine to medium sand. All British coasts
Glycera rouxi Audouin and Milne Edwards (II.46)

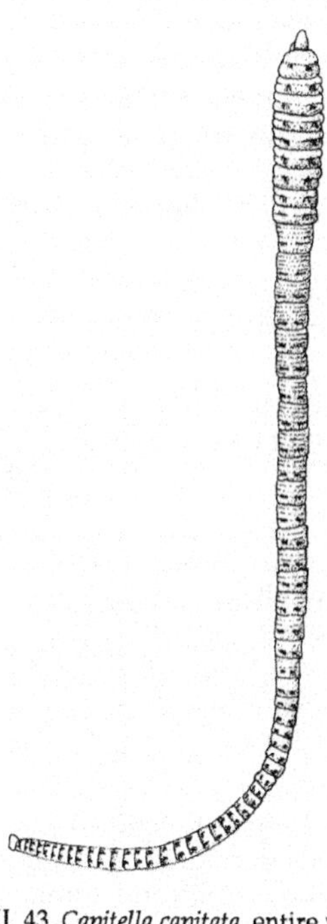
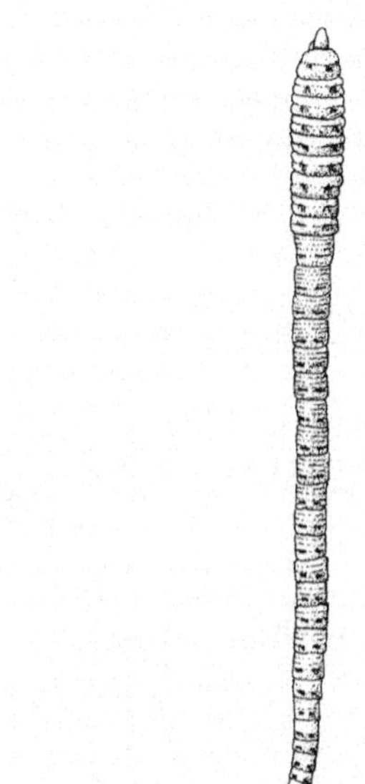

II. 43 *Capitella capitata*, entire worm

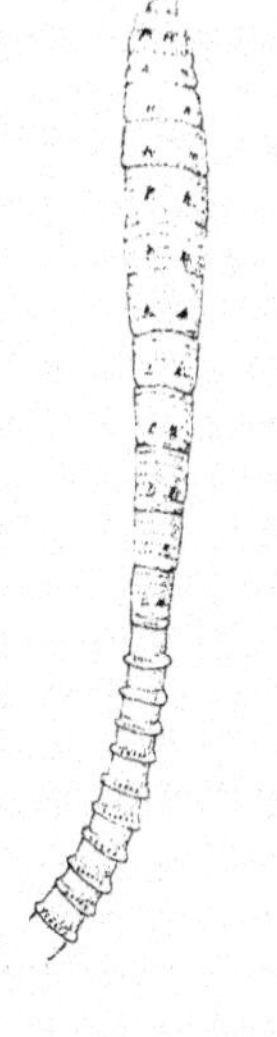

II. 44 *Notomastus latericeus*, broad front region and part of tail

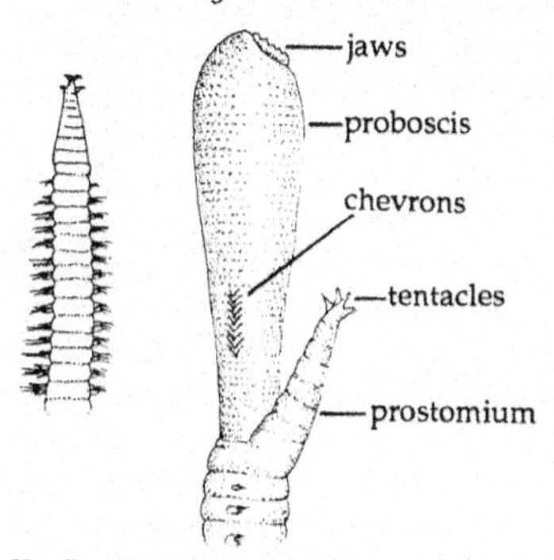

II. 45 *Goniada maculata*, front end of worm (left) and head with proboscis extended (right)

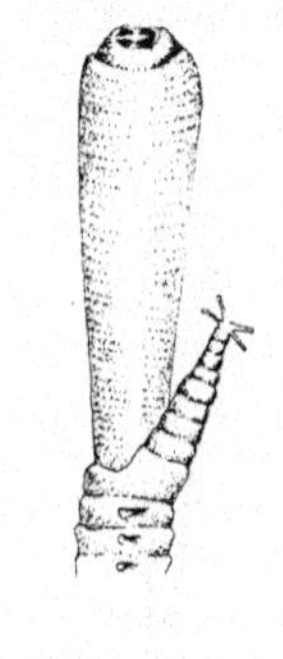

II. 46 *Glycera rouxi*, head end with proboscis extended

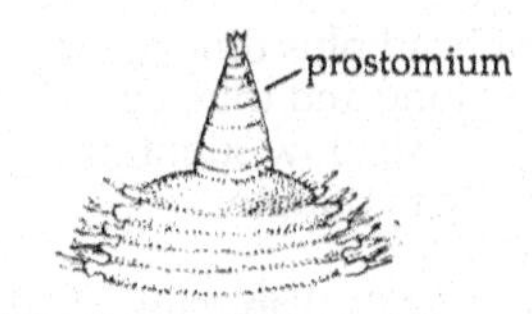

II. 47 *Glycera alba*, head end

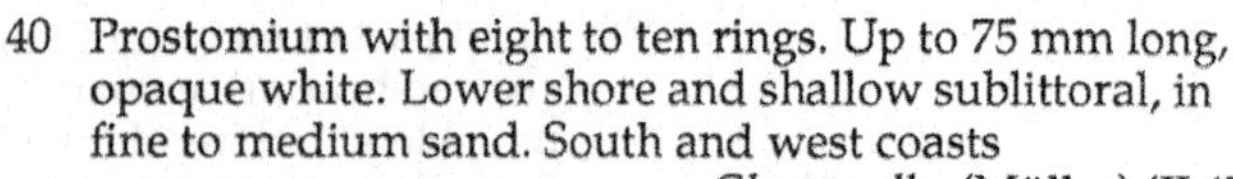

40 Prostomium with eight to ten rings. Up to 75 mm long, opaque white. Lower shore and shallow sublittoral, in fine to medium sand. South and west coasts
Glycera alba (Müller) (II.47)

– Prostomium with 14–18 rings. Up to 100 mm long; pink, iridescent. From MTL into shallow sublittoral, in fine to medium sand. South and west coasts
Glycera tridactyla Schmarda (II.48)

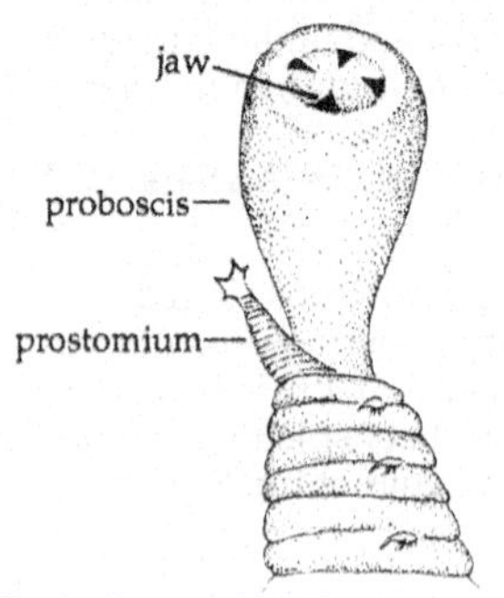

II. 48 *Glycera tridactyla*, head with proboscis extended

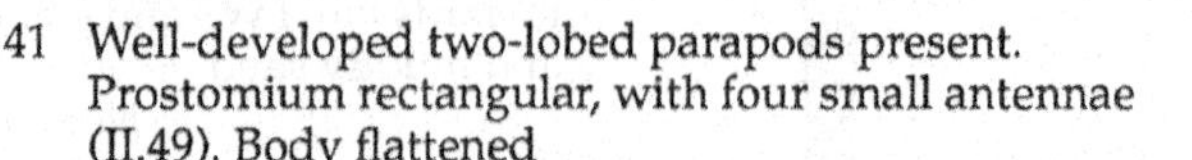

41 Well-developed two-lobed parapods present. Prostomium rectangular, with four small antennae (II.49). Body flattened 42

– Parapods small, asymmetrical, with small notopod and larger neuropod (II.54). Prostomium conical, with no antennae 44

42 Proboscis with numerous rows of papillae, all of similar size. Up to 250 mm long; opaque white, lustrous. Intertidal. Common, and often abundant, on all British coasts *Nephtys caeca* Fabricius (II.49,50)

– Proboscis with rows of papillae, of different sizes, including one large middle one (II.51) 43

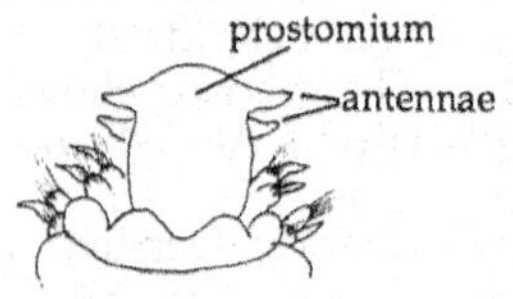

II. 49 *Nephtys caeca*, head end

43 Branchial cirri of rear parapods as long as the gill (II.52). Body oval in cross section, up to 100 mm long; opaque, lustrous white. Intertidal and shallow sublittoral, often abundant. All British coasts
Nephtys cirrosa Ehlers (II.51,52)

– Branchial cirri of rear parapods much shorter than gills. Body square in cross section, up to 200 mm long; opaque white, pearly. Intertidal and shallow sublittoral. Common on all British coasts
Nephtys hombergi Savigny (II.53)

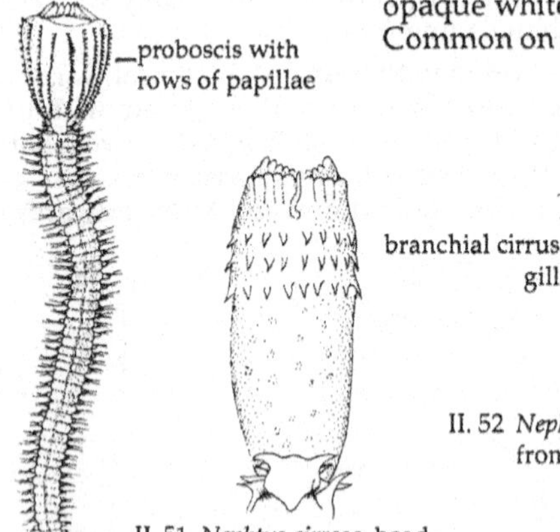

II. 50 *Nephtys caeca*, entire worm with proboscis extended

II. 51 *Nephtys cirrosa*, head with proboscis extended

branchial cirrus
gill

II. 52 *Nephtys cirrosa*, parapod from rear end of worm

notopod
branchial cirrus
gill
neuropod
ventral cirrus

II. 53 *Nephtys hombergi*, parapod from rear end of worm

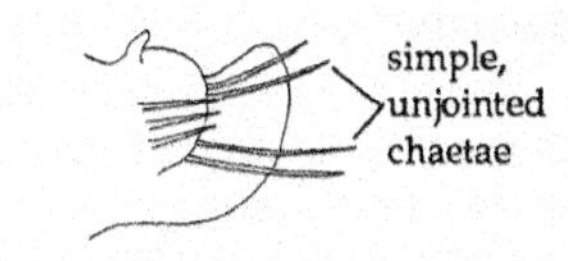

II. 54 *Arabella iricolor*, parapod

44 Parapods with simple chaetae only (II.54) — 45

– Some parapods with hooded, crotchet chaetae (II.55) jointed in the first 20–30 segments, unjointed thereafter. Up to 300 mm long; pink, orange or brown, iridescent. At LWS and sublittoral, in muddy sand, especially under stones and among eelgrass. All British coasts
Lumbrineris latreilli Audouin and Milne Edwards (II.56)

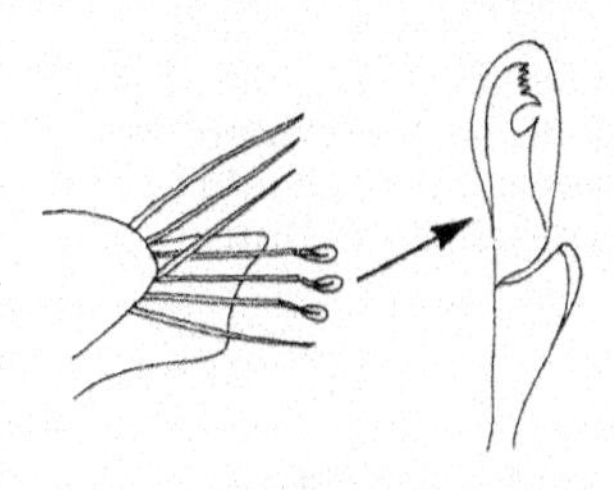

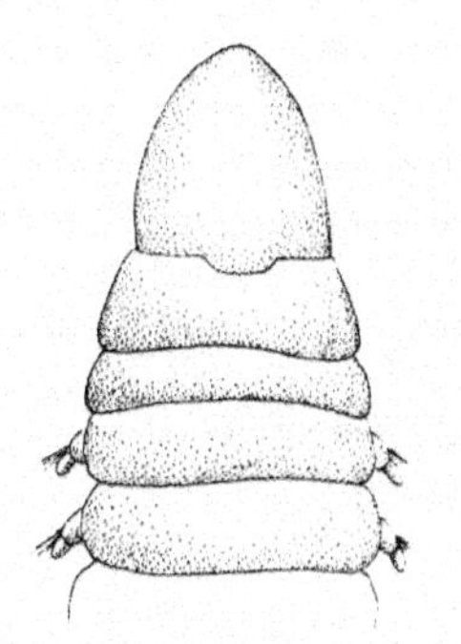

II. 55 *Lumbrineris latreilli*, parapod

45 Ventral chaetae of each parapod include one large, blunt chaeta (II.57). Body slender, up to 160 mm long; yellow, pink or greenish, iridescent. At LWS and sublittoral, in fine muddy sand. South and west coasts
Drilonereis filum Claparède (II.58)

– Ventral chaetae of each parapod all similar in size. Body cylindrical, tapered at each end, up to 600 mm long; pinkish grey, iridescent. At LWS and shallow sublittoral, in fine muddy sand. South west coasts only
Arabella iricolor (Montagu) (II.59,60)

II. 56 *Lumbrineris latreilli*, head

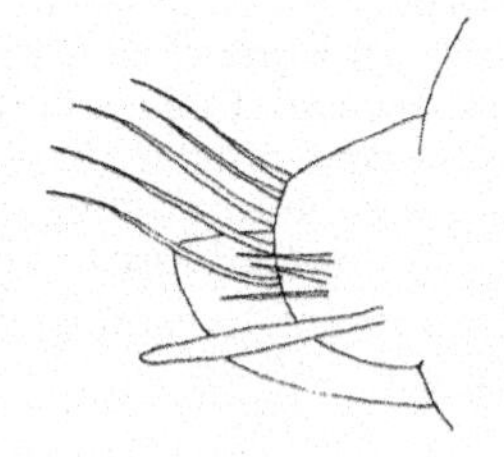

II. 57 *Drilonereis filum*, parapodium (after George and Hartmann-Schröder)

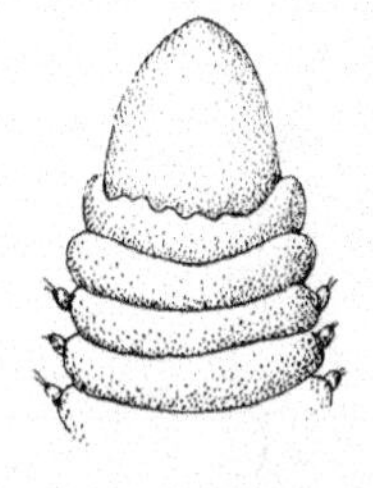

II. 58 *Drilonereis filum*, head and first few segments (after George and Hartmann-Schröder)

II. 59 *Arabella iricolor*, head

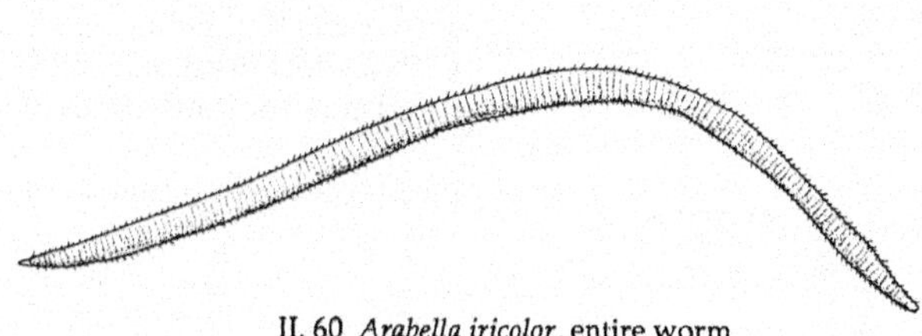

II. 60 *Arabella iricolor*, entire worm

Key III. Cumaceans

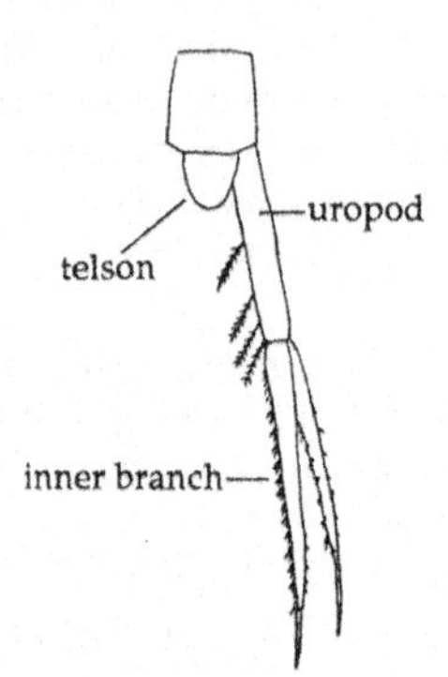

III.1 *Pseudocuma longicornis*, telson and right uropod

Cumaceans are usually collected by sieving, or night-time plankton sampling, and are easily recognized among other small crustaceans by their tadpole-like shape (fig. 42). The sexes are different; in females the second antenna is very short, and there are no pleopods. Both the shape and the surface sculpture of the carapace are also liable to differ between males and females. The following key will allow the identification of the six most common species; specimens which do not key out here should be identified using Jones (1976), who describes all British species.

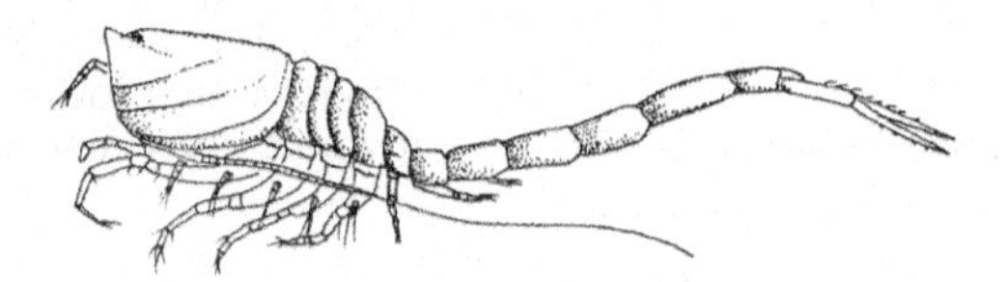

III. 2 *Pseudocuma longicornis*, male

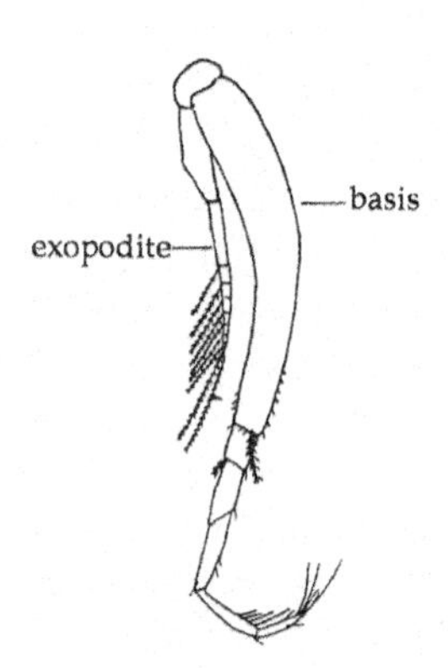

III. 3 *Iphinoe trispinosa*, pereopod 1

1 A small rounded telson present at hind end of body (III.1); inner branch of uropod with only one segment. Carapace with three oblique folds on each side. Up to 4 mm long, with dark brown markings. Intertidal and shallow sublittoral, often in brackish water. All British coasts, common *Pseudocuma longicornis* (Bate) (III.1,III.2)

– Telson absent, or completely fused with last body segment 2

2 First pereopod with a small exopodite (III.3); remaining pereopods unbranched, without exopodites 3

– Exopodites present on pereopods 1–3 or 1–4, though sometimes very small (III.4) 5

3 Only four limb-bearing thoracic segments visible from above, the first being fused with the carapace. Carapace with a short horn on each side. Up to 7 mm long. Intertidal, presently known only from Channel Isles *Eocuma dollfusi* Calman (III.5)

– Five limb-bearing thoracic segments visible from above 4

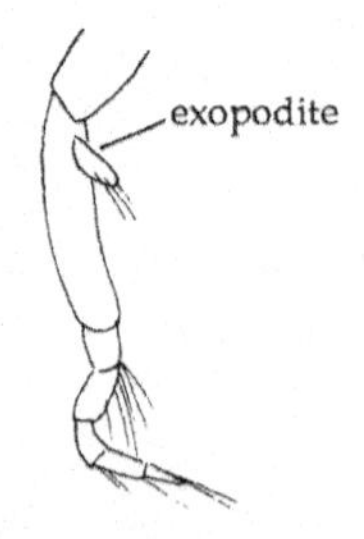

III. 4 *Cumopsis goodsiri*, pereopod 2

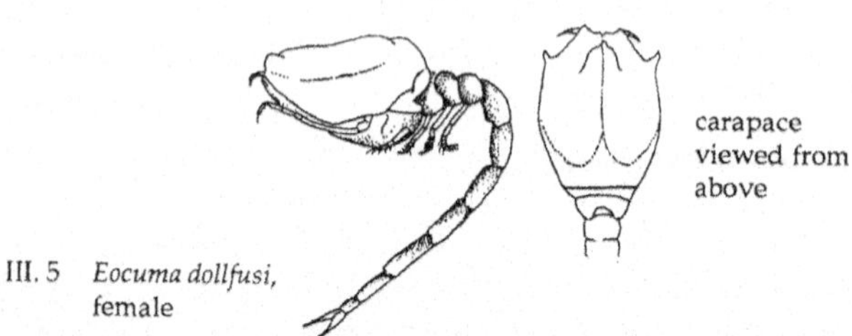

III. 5 *Eocuma dollfusi*, female

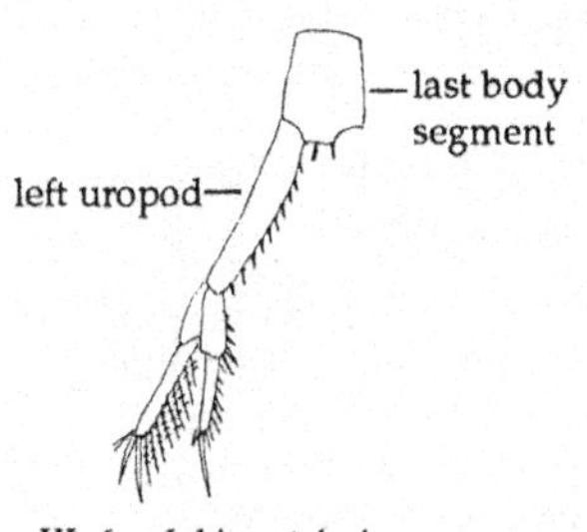

III. 6 *Iphinoe trispinosa*

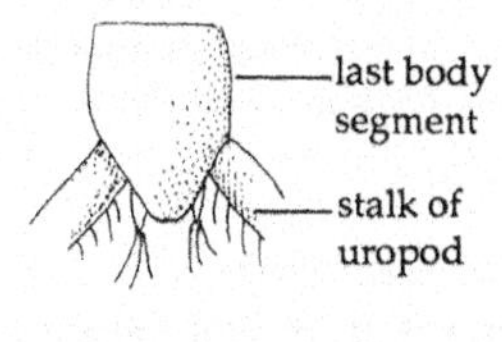

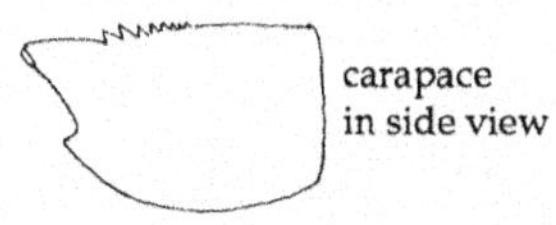

III. 8 *Iphinoe tenella*

4 First pereopod with the basis longer than the remaining segments together (III.3). Two short spines on hind edge of last abdominal segment (III.6). Up to 10 mm long, white to pale yellowish. In fine sands, intertidal and sublittoral. All British coasts
Iphinoe trispinosa (Goodsir) (III.7)

– First pereopod with shorter basis. Last body segment with six spines on hind edge. Up to 8 mm long. On muddy sand, intertidal and shallow sublittoral. South coasts only *Iphinoe tenella* Sars (III.8)

5 Carapace with two folds on each side. Up to 6 mm long, with purplish brown patches on carapace and pleon segment 5. In fine sand, from MTL into shallow sublittoral. All British coasts
Cumopsis goodsiri (Van Beneden) (III.4,9)

– Carapace without folds on the sides. Inner branch of uropod with spines along the side and a thin spine at the tip. Up to 6 mm long. Reported rarely from south coasts of England *Cumopsis longipes* (Dohrn) (III.10)

[*C. fagei* Bacescu occurs on exposed sandy beaches in the Channel Isles. It is distinguished from *C. longipes* by its lack of spines on the edges of the inner branch of the uropod.]

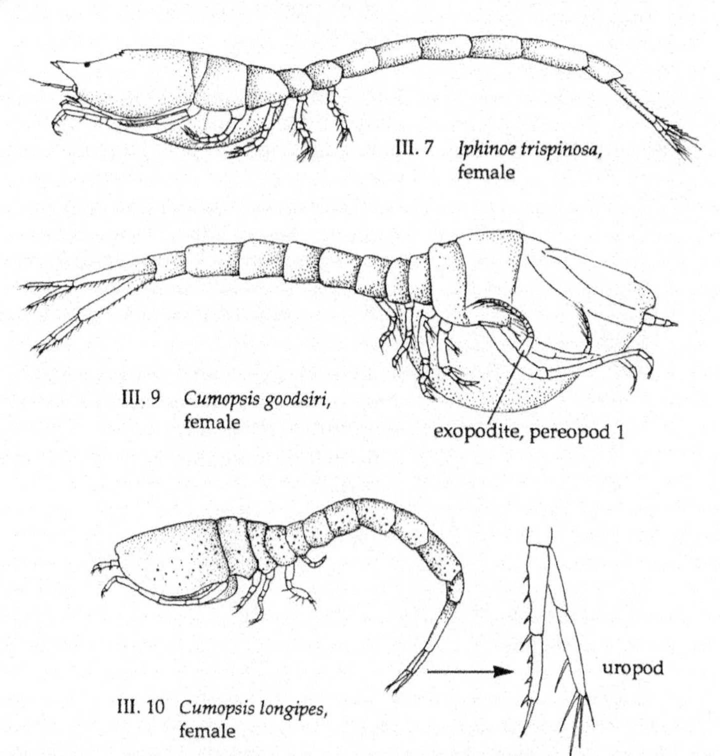

III. 7 *Iphinoe trispinosa*, female

III. 9 *Cumopsis goodsiri*, female

III. 10 *Cumopsis longipes*, female

Key IV. Mysids

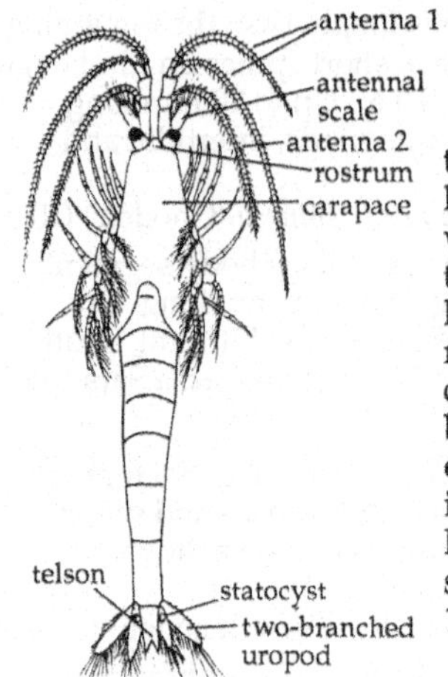

IV. 1 *Gastrosaccus sanctus*, male

Mysids are shrimp-like animals with elongate, transparent bodies and feathery, two-branched thoracic limbs. The eyes are large and prominent; the carapace, which may have a rostrum (IV.1) at the front, is not attached to the body at its hind edge. Antenna 2 bears a flat, plate-like antennal scale (IV.1) which is important in species recognition. The tailfan – also important in identification – consists of an elongate telson and a pair of long, two-branched uropods. The inner branch of each uropod – the endouropod – has a conspicuous, rounded statocyst close to its base. All British species are described in a Ray Society Monograph (Tattersall and Tattersall, 1951), and all coastal species may be identified using a *Field Studies* key (Makings 1977).

1 Telson distinctly notched or cleft at tip (IV.1) — 2
– Telson rounded, pointed or straight at tip, but without a cleft or notch — 8

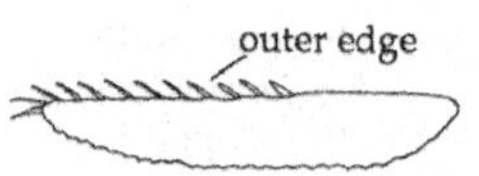

IV. 2 *Gastrosaccus sanctus*, outer branch of left uropod

2 Outer branch of uropod with stout spines on its outer edge (IV.2) — 3
– Outer branch of uropod with long bristles on its outer edge (IV.3) — 4

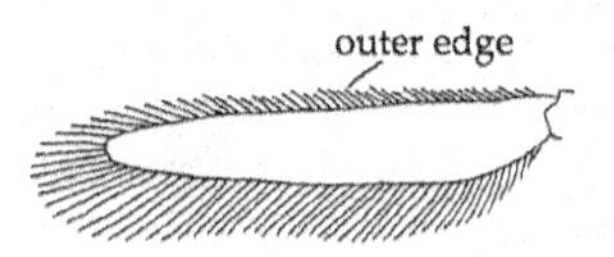

IV. 3 *Praunus inermis*, outer branch of left uropod

3 Fifth abdominal segment narrow when viewed from above, ridged along its length, with a stout spine on its hind edge. Up to 21 mm long. Common on all British coasts *Gastrosaccus spinifer* (Goes) (IV.4)
– Fifth abdominal segment narrow, but not ridged and without a spine. Telson twice as long as broad, with five or six spines on each side. Up to 15 mm long. In mud or muddy sand, shallow sublittoral. South and west coasts only *Gastrosaccus sanctus* (Van Beneden) (IV.1,IV.2)

4 Antennal scale slender, parallel sided, with a short anterior lobe which only just projects beyond the subterminal spine (IV.5) — 5
– Antennal scale slender or oval, but with a long anterior lobe (IV.6) — 6

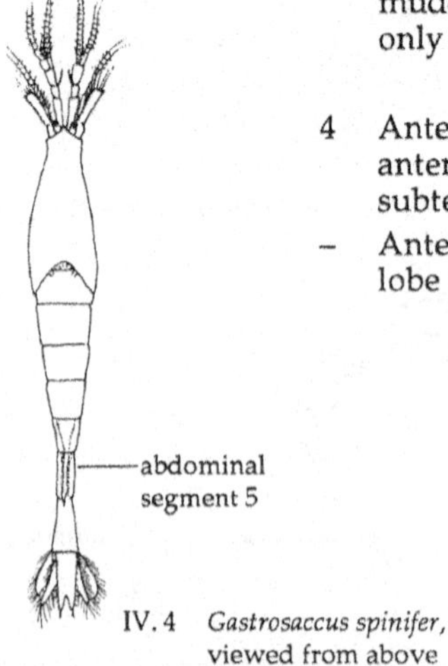

IV. 4 *Gastrosaccus spinifer*, viewed from above (with legs removed)

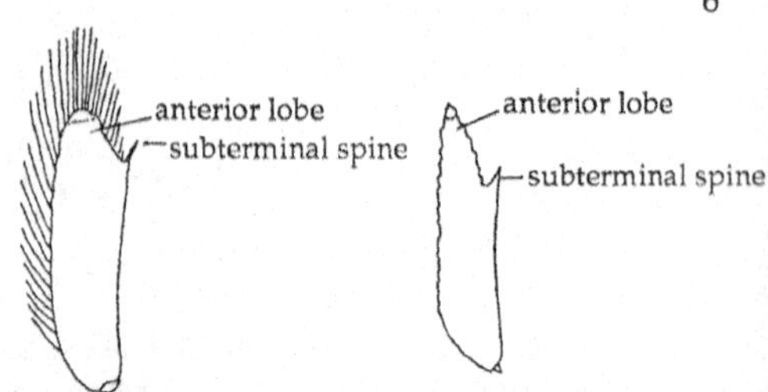

IV. 5 *Praunus flexuosus*, right antennal scale

IV. 6 *Schistomysis spiritus*, right antennal scale

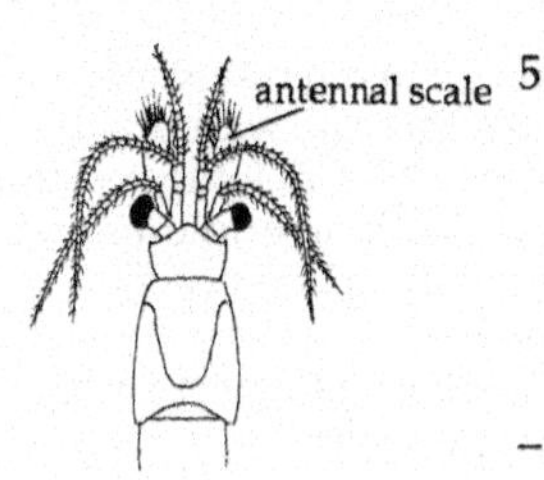

IV. 7 *Praunus inermis*, head and thorax from above

5 Antennal scale four times as long as broad. Eyestalk wider than long. Telson with 15–17 spines on each side. Up to 15 mm long; pale reddish brown, underside of thorax with two pairs of black spots. Lower shore, especially among drifting weed in shallow water. All British coasts, common
Praunus inermis (Rathke) (IV.7; pl.2.7)

– Antennal scale seven or eight times as long as broad. Eyestalk as long as, or longer than, wide. Telson with 22–28 spines on each side. Up to 24 mm long; translucent yellow to dark grey, with eight pairs of black spots on underside of thorax. Lower shore, among weeds, sometimes in brackish water. Common on all British coasts
Chameleon Shrimp *Praunus flexuosus* (Müller) (IV.8)

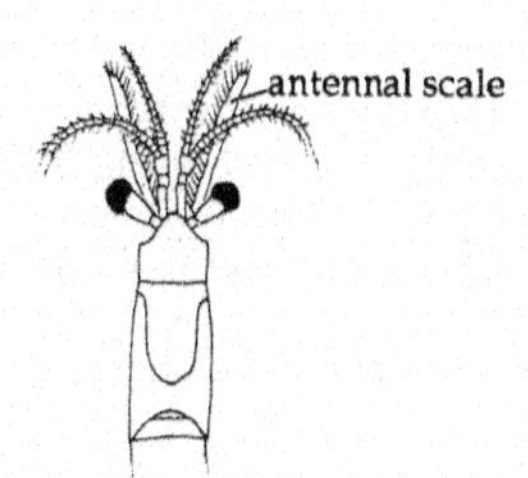

IV. 8 *Praunus flexuosus*, head and thorax from above

6 Eyes projecting well beyond edge of carapace, eyestalk longer than wide. Tip of inner branch of uropod slightly curved (IV.9). Up to 18 mm long; transparent, sometimes with pale reddish brown markings. Abundant on all British coasts
Ghost Shrimp *Schistomysis spiritus* (Norman) (IV.10)

– Eyes scarcely protruding beyond edge of carapace, with short broad eyestalks. Tip of inner branch of uropod straight 7

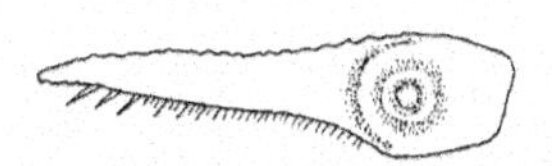

IV. 9 *Schistomysis spiritus*, inner branch of left uropod

7 Antennal scale elongate, more than three times as long as broad. Telson with 26 evenly spaced, short spines around edges. Up to 19 mm long, more opaque than *S. spiritus*. Mostly sublittoral, infrequent on lower shore. Common on all British coasts
Schistomysis ornata (G.O. Sars) (IV.11)

– Antennal scale oval, less than three times as long as broad (IV.12). Telson deeply cleft, with 17–23 spines. Up to 10 mm long, translucent. On lower shore and in shallow sublittoral. South and west coasts only
Paramysis arenosa (G.O. Sars) (IV.13)

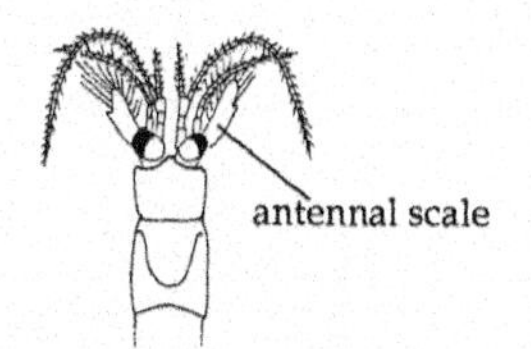

IV. 11 *Schistomysis ornata*, head and thorax viewed from above

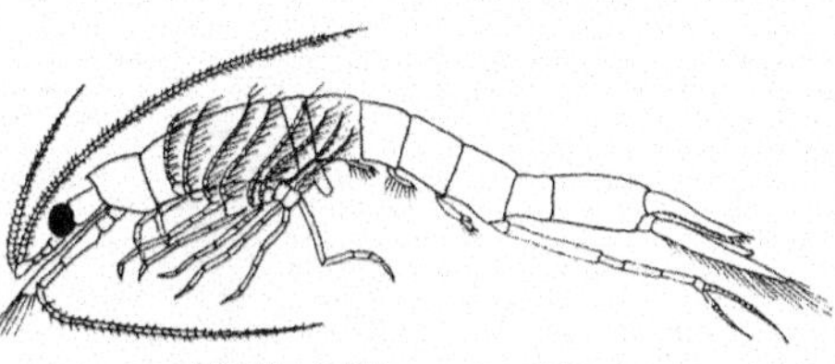

IV. 10 *Schistomysis spiritus*, male

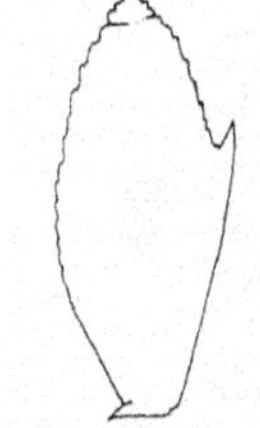

IV. 12 *Paramysis arenosa*, right antennal scale

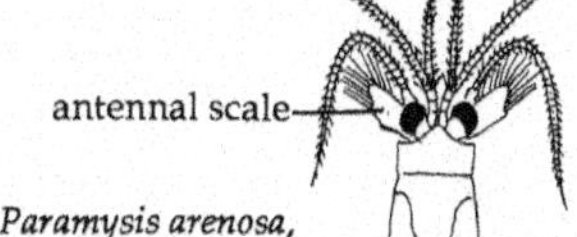

IV. 13 *Paramysis arenosa*, head and thorax viewed from above

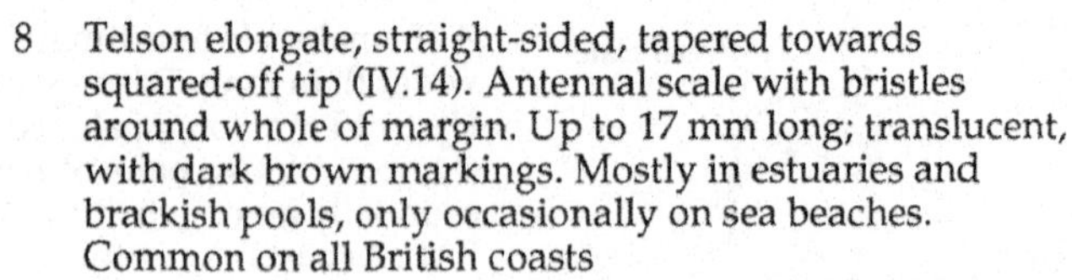

8 Telson elongate, straight-sided, tapered towards squared-off tip (IV.14). Antennal scale with bristles around whole of margin. Up to 17 mm long; translucent, with dark brown markings. Mostly in estuaries and brackish pools, only occasionally on sea beaches. Common on all British coasts
Neomysis integer (Leach) (IV.15)

– Telson with rounded tip (IV.16,IV.18) 9

9 Telson slender, tapered (IV.16). Carapace with very prominent rostrum. Up to 22 mm long. Lower shore and shallow sublittoral. Common on all British coasts
Siriella armata (Milne Edwards) (IV.17)

– Telson with broadly rounded tip (IV.18). Carapace with very short rostrum, reaching only halfway along first basal segment of antenna 1. Up to 17 mm long, pale reddish. Lower shore and shallow sublittoral. Common on all British coasts
Leptomysis lingvura (G.O. Sars) (IV.19)

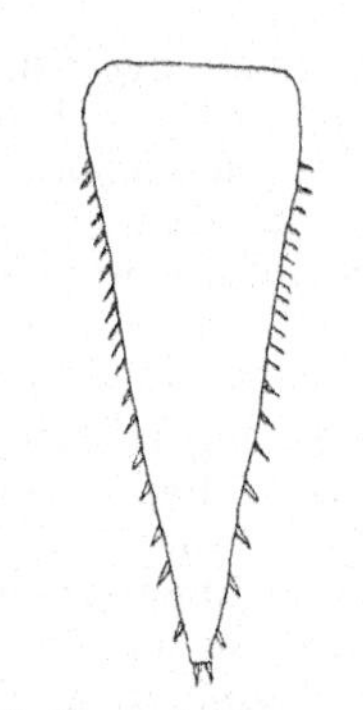

IV. 14 *Neomysis integer*, telson

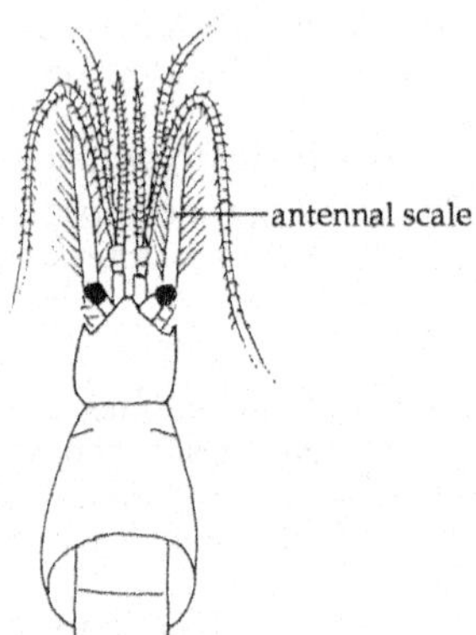

IV. 15 *Neomysis integer*, head and thorax viewed from above

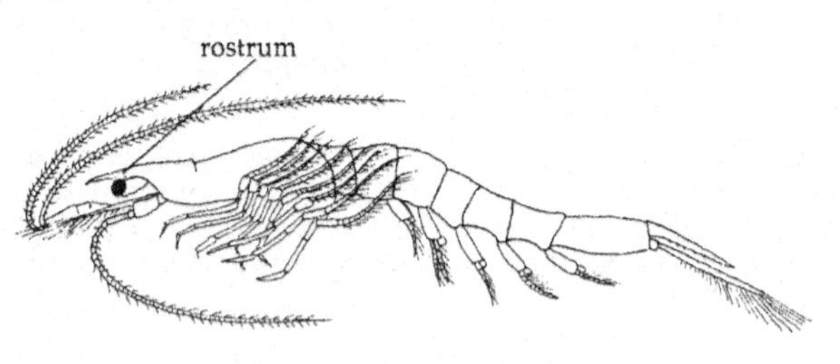

IV. 17 *Siriella armata*, male

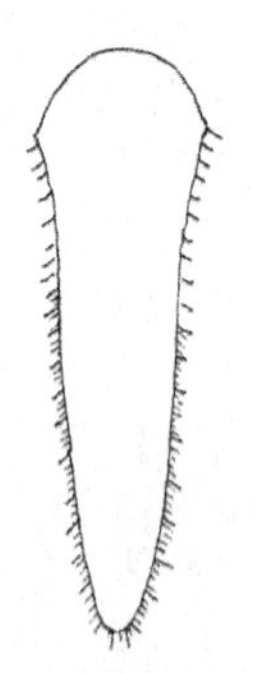

IV. 16 *Siriella armata*, telson

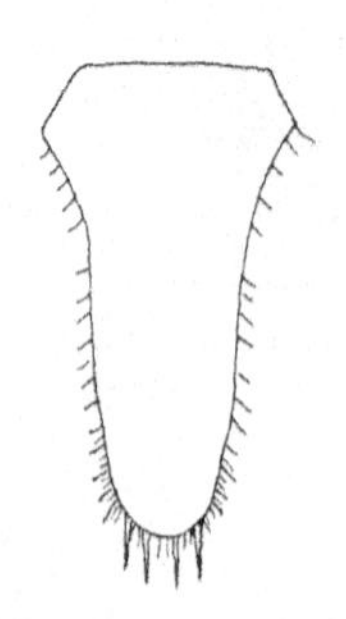

IV. 18 *Leptomysis lingvura*, telson

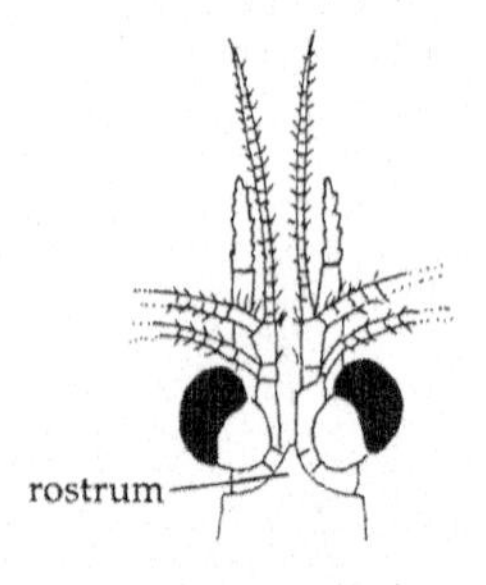

IV. 19 *Leptomysis lingvura*, head

Key V. Isopods

British coastal isopods have been described by Naylor (1972, 1990). The following key includes three resident infaunal species, together with a number of others commonly found stranded at LWM, and a common supralittoral species. Figures V.1–V.9 are based on Naylor (1972).

1 Uropods short, cylindrical, at hind end of body projecting from a broad, flattened, pleotelson. Body oval; males up to 2.5 mm long, females to 5.0 mm. At HWM, in damp sand beneath stones, usually in brackish water areas — *Jaera albifrons* Leach (V.1)

– Uropods flattened, not cylindrical; on the underside of the pleotelson, or attached to its sides — 2

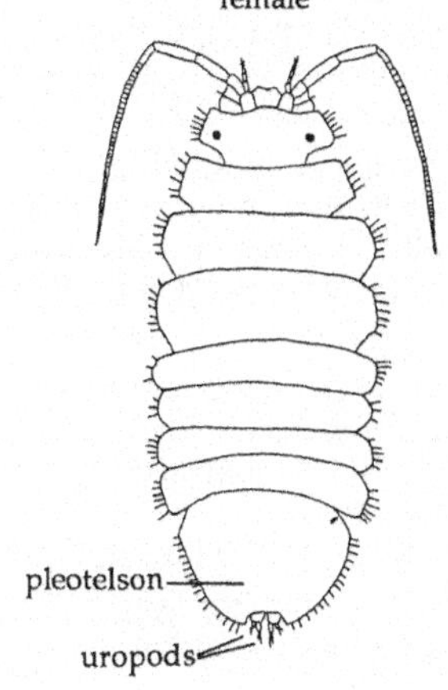

V. 1 *Jaera albifrons*

2 Uropods forming flat plates on the underside of the pleotelson (V.2) — 3

– Uropods short and flattened, attached on each side at end of body (V.7) — 6

3 Hind edge of pleotelson concave (V.2) — 4

– Hind edge of pleotelson pointed or rounded — 5

4 Coxal plates more prominent, those of rear body segments reaching back as far as hind edge of segment. Males up to 30 mm long, brown with white markings; females to 18 mm, dark brown with white banding. Often stranded on lower shore on drift algae. All British coasts — *Idotea emarginata* (Fabricius) (V.2)

– Body very elongate; coxal plates small, none projecting backwards to reach hind edge of body segment to which they are attached. Up to 40 mm long, green or brown with dark or light longitudinal stripes. Sublittoral, sometimes stranded on lower shore, more often swimming at water's edge at LWS. South and west coasts only — *Idotea linearis* (L.) (V.3)

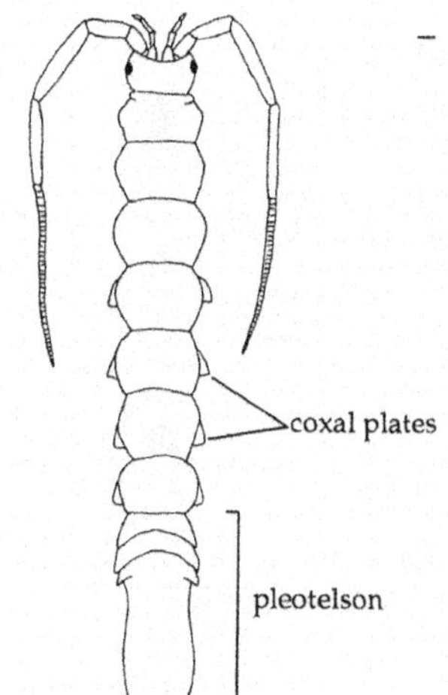

V. 3 *Idotea linearis*

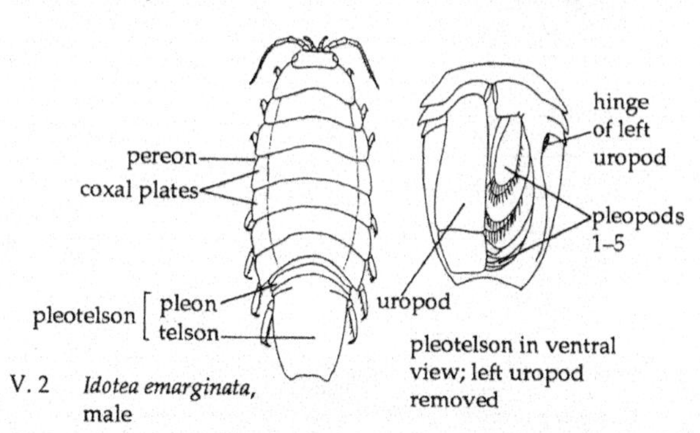

V. 2 *Idotea emarginata*, male

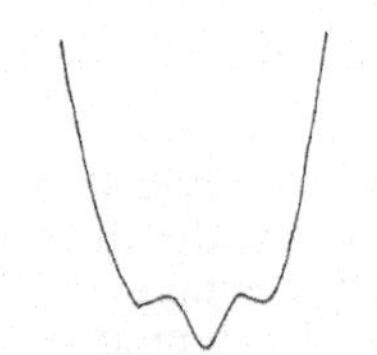

V. 4 *Idotea baltica*, outline of pleotelson

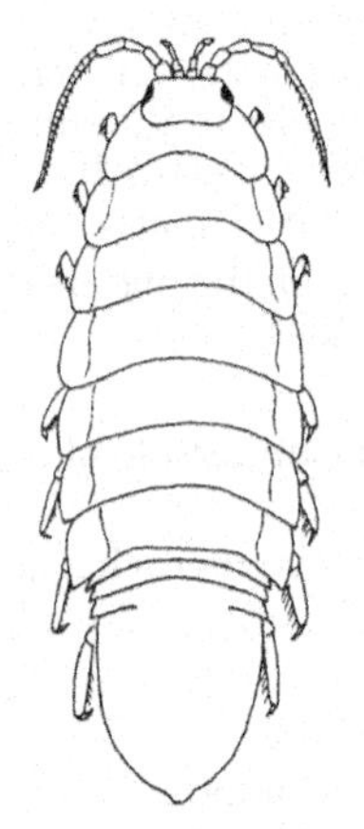

V. 5 *Idotea neglecta*, male

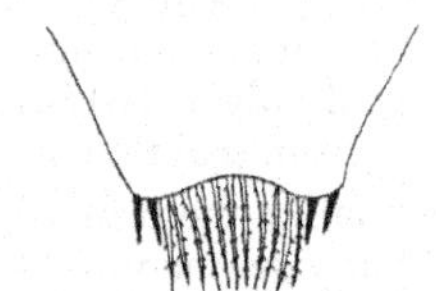

V. 6 *Eurydice spinigera*, hind edge of pleotelson

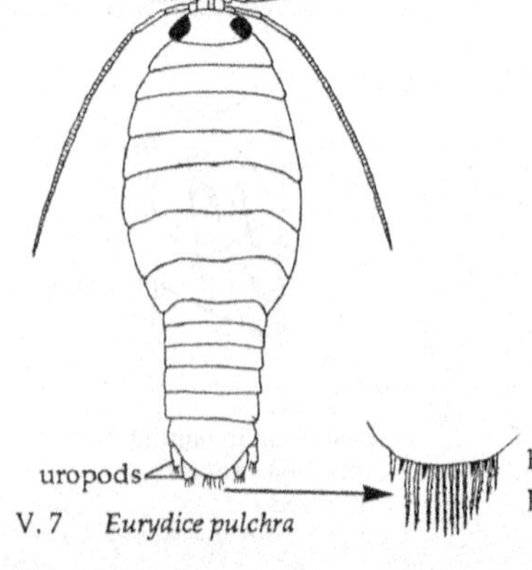

V. 7 *Eurydice pulchra*

5 Hind edge of pleotelson more or less three-pointed. Males up to 30 mm long, females to 18 mm; green or brown, often with white spots and lines
Idotea baltica (Pallas) (V.4)

– Pleotelson rounded at end, with a short, pointed tip. Males up to 30 mm long, females to 16 mm; brown with white stripes or marbling. Sublittoral, often on lower shore on drift algae *Idotea neglecta* Sars (V.5)

6 Hind edge of pleotelson concave, with two stout spines at each corner, and feathery bristles between. Up to 9 mm long. At LWST and sublittoral, often in surf plankton. South and west coasts only
Eurydice spinigera Hansen (V.6)

– Hind edge of pleotelson broadly rounded; two short spines on each side, hidden among feathery bristles (V.7) 7

7. Coxal plates on pereon segment 6 extending sharply backwards (V.8). Black spots on upper and lower surfaces, and along sides. Males up to 8 mm long, females to 6.5 mm. Intertidal, swimming on rising tides. All British coasts, common
Eurydice pulchra Leach (pl. 2.1; V.7)

– Coxal plates on pereon segment 6 not extending sharply backwards. Black spots on upper surface only. Smaller than preceding species. Intertidal; uncommon, reported from North Wales and Bristol Channel
Eurydice affinis Hansen (V.9)

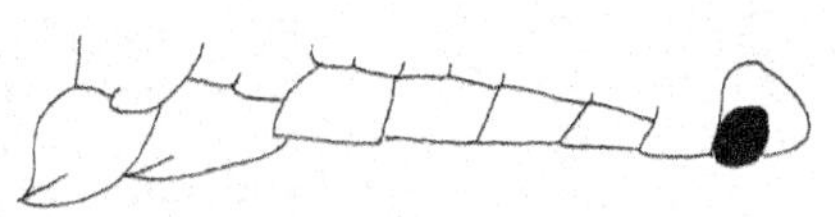

V. 8 *Eurydice pulchra*, head and coxal plates in side view

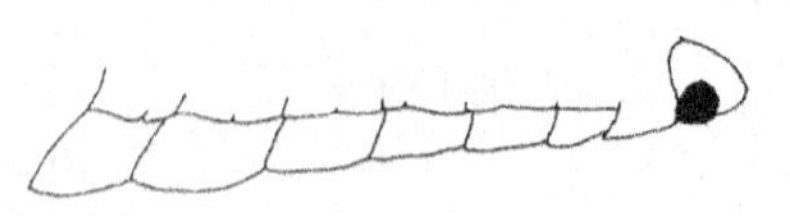

V. 9 *Eurydice affinis*, head and coxal plates in side view

Key VI. Amphipods

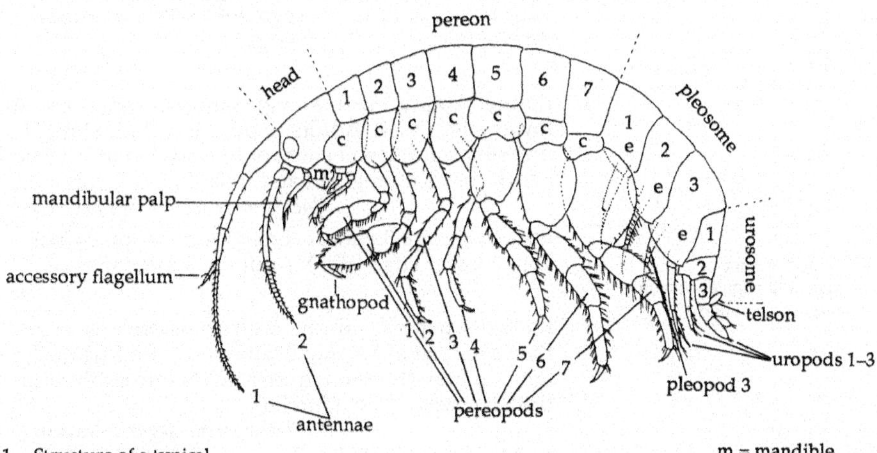

VI. 1 Structure of a typical gammaridean amphipod (after Lincoln, 1979)

m = mandible
c = coxal plate
e = epimeral plate

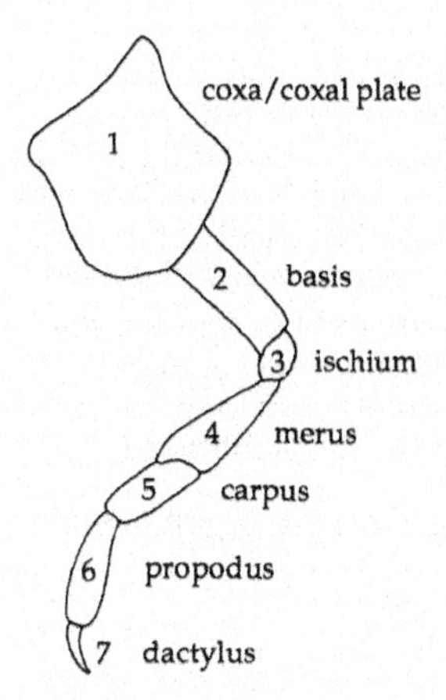

VI. 2 The segments (articles) of an amphipod limb

Identification of amphipods requires a binocular microscope and considerable patience. Specimens are best examined in watch glasses of water, having first been preserved in 70% alcohol. Substage illumination is important to see fine details, such as spines and bristles, and a pair of seekers is necessary to tease out the different limbs. Identification need not be difficult; each species is defined by discrete morphological features, but it is necessary to be familiar with the terms used in amphipod anatomy. These are all given in Figures VI.1–2. Sexes are often different; the male, for example, may have larger gnathopods and longer antennae than the female, which is recognised by the plates of the egg-brooding chamber, between the front series of legs. There are more than 250 species of Amphipoda around Britain's coast, all of which are described and illustrated by Lincoln (1979). It must be stressed that the following key will permit identification of resident infaunal species *only*, together with a few common migrants. If a particular specimen does not correspond exactly with key characters, and of course illustrations, given here then recourse must be had to Lincoln's (1979) comprehensive keys.

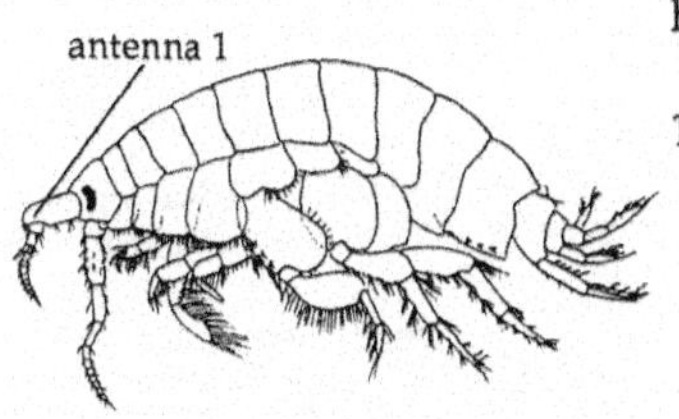

VI. 3 *Bathyporeia guilliamsoniana*, female

1 First segment of antenna 1 prominent, almost as large as head (VI.3); remaining segments attached to its lower surface, the antenna thus right-angled (geniculate). Antenna 2 longer than 1, more than half body length in males. Generally plump animals with inconspicuous gnathopods, and pereopods broadened for digging — 2

– Antenna 1 long or short, stout or slender, but never geniculate — 7

2 Top surface of urosome segment 1 with forwardly directed bristles and backwardly directed spines (VI.4) 3
– Top surface of urosome segment 1 with forwardly directed bristles only, no spines 6

3 Epimeral plate 3 with a single group of spines just above lower edge (VI.4). Adults small, about 3 mm long. In fine silty sand; lower shore only, MLWS into sublittoral. South coasts of England and Ireland, locally abundant *Bathyporeia nana* (Toulmond) (VI.4)
– Epimeral plate 3 with several groups of spines just above lower edge (VI.5). Adults 5–8 mm long 4

4 Epimeral plate 3 with smoothly rounded hind edge. First segment of antenna 1 rounded at tip. Adults 5 mm long. In fine silty sand, from MLWN into sublittoral. All British coasts, often abundant *Bathyporeia elegans* Watkin (VI.5)
– Epimeral plate 3 with distinct tooth on lower hind corner (sometimes worn in older animals, but still appearing uneven) (VI.6) 5

5 Epimeral plate 3 with well developed tooth. Coxae 2 and 3 each with similar tooth on lower hind corner (VI.6). First segment of antenna 1 rounded at tip. Adults 8 mm long. In fine and medium sand, from MLWN into shallow sublittoral. All British coasts, locally abundant *Bathyporeia guilliamsoniana* (Bate) (VI.3)
– Epimeral plate 3 with small tooth only. Coxae 2 and 3 without tooth on lower hind corner. First segment of antenna 1 with angular tip. Adults 6 mm long. In fine to medium sand, from MTL into shallow sublittoral. All British coasts, locally abundant *Bathyporeia pelagica* (Bate) (VI.7)

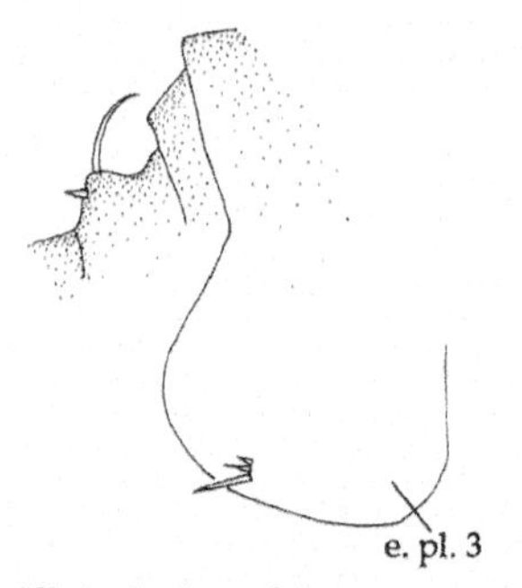

VI. 4 *Bathyporeia nana*, male, top edge of urosome, and epimeral plate 3

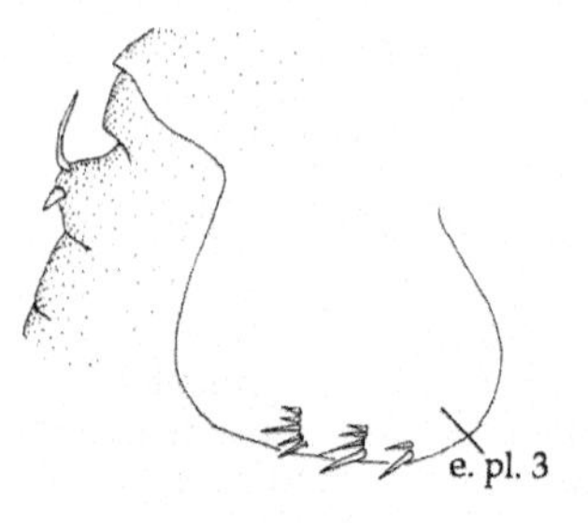

VI. 5 *Bathyporeia elegans*, female, top edge of urosome, and epimeral plate 3

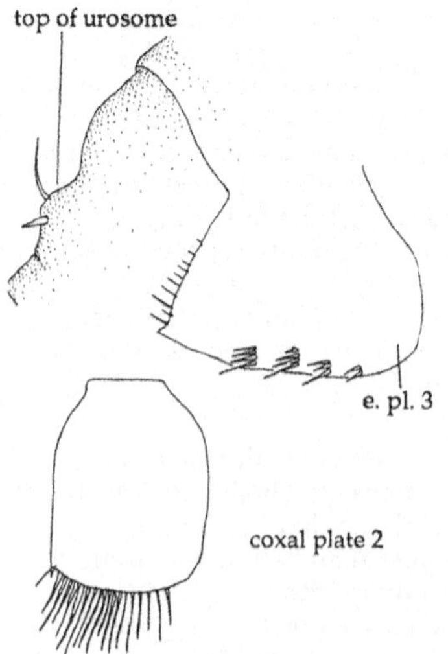

VI. 6 *Bathyporeia guilliamsoniana*, female

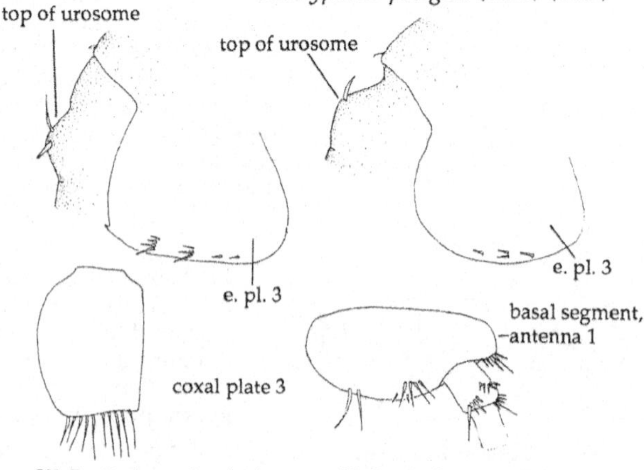

VI. 7 *Bathyporeia pelagica*, female

VI. 8 *Bathyporeia pilosa*, female

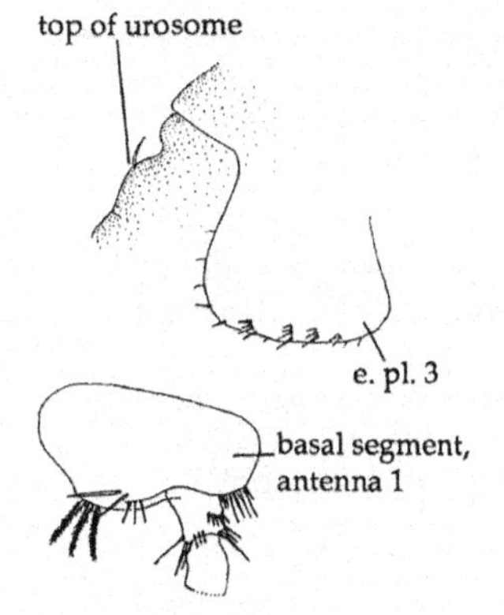

VI. 9 *Bathyporeia sarsi*, female

6 Epimeral plate 3 with two or three groups of spines above lower edge (VI.8). First segment of antenna 1 with tapered, rounded tip. Adults 6 mm long. Entirely intertidal, from MHWN to ELWS, in fine to medium sand. Common on all British coasts, tolerant of reduced salinity, often extremely abundant *Bathyporeia pilosa* Lindström (pl. 2.3;VI.8)

– Epimeral plate 3 with four to six groups of spines above lower edge. First segment of antenna 1 with broadly rounded tip. Adults 8 mm long. In fine to medium sand, from MHWN to ELWS, most common below *B. pilosa* zone and above *B. pelagica* zone. South coasts only, locally abundant *Bathyporeia sarsi* Watkin (VI.9)

7 Plump-bodied animals with pereopods broad or flattened (pl. 2.2,9) – modified for digging – and fringed with long bristles 8

– Slender animals without broadened pereopods 11

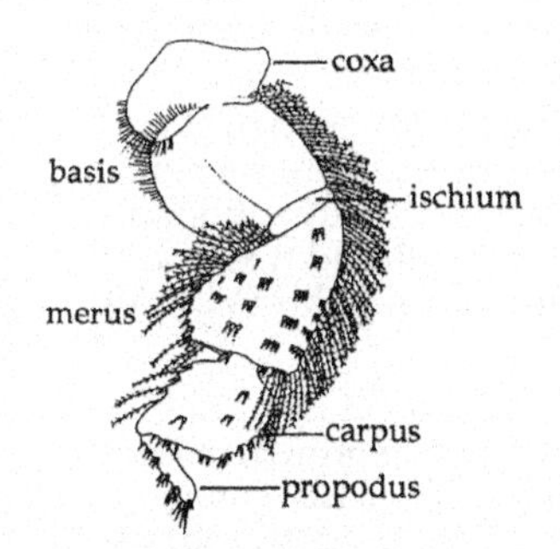

VI. 10 *Haustorius arenarius*, female, pereopod 5

8 Pereopods 3–7 each with five segments only (VI.10), no dactylus. Adults 13 mm long, white. Large active animals with especially broad pereopods. Upper shore only, common on all coasts *Haustorius arenarius* (Slabber) (pl. 2.9)

– Each pereopod with six segments 9

9 Uropod 1 with smooth, strongly curved branches (VI.11). Adults 8 mm long, white. At LWS only. Common on all British coasts *Urothoe marina* (Bate) (pl. 2.2)

– Uropod 1 with straight branches, often with spines or bristles 10

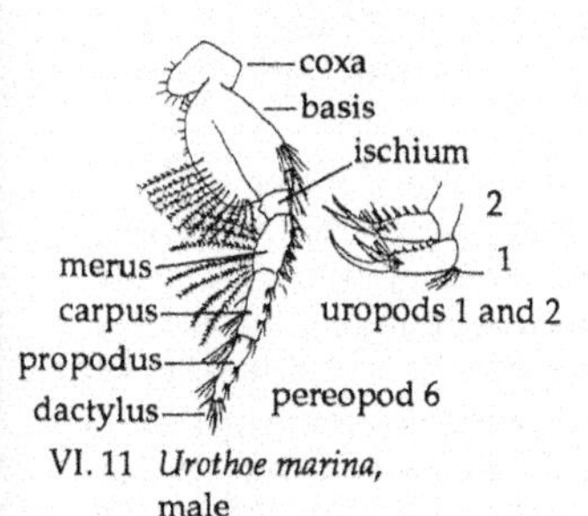

VI. 11 *Urothoe marina*, male

10 Pereopod 5 with carpus about as wide as long (VI.12), scarcely wider than preceding merus. Adults 8 mm long. In fine to medium sand, from MTL to sublittoral. Not common *Urothoe brevicornis* Bate (VI.12)

[*U. elegans* Bate is similar, with pereopod 5 carpus and merus of equal length, but is distinguished by its shorter accessory flagellum. It is strictly sublittoral.]

– Pereopod 5 with carpus more than twice as wide as merus (VI.13). Adults 6 mm long, white. Lower shore and shallow sublittoral. Southern coasts only, from The Wash to the Bristol Channel *Urothoe poseidonis* Reibisch (VI.13)

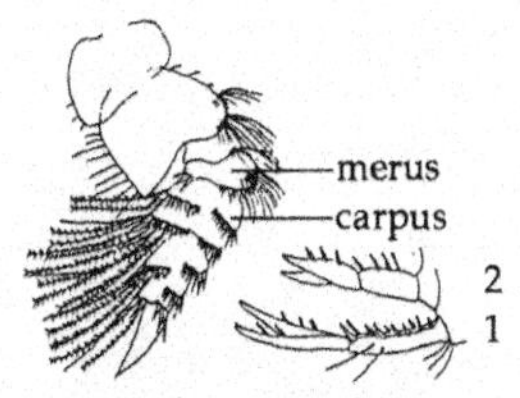

VI. 12 *Urothoe brevicornis*, female

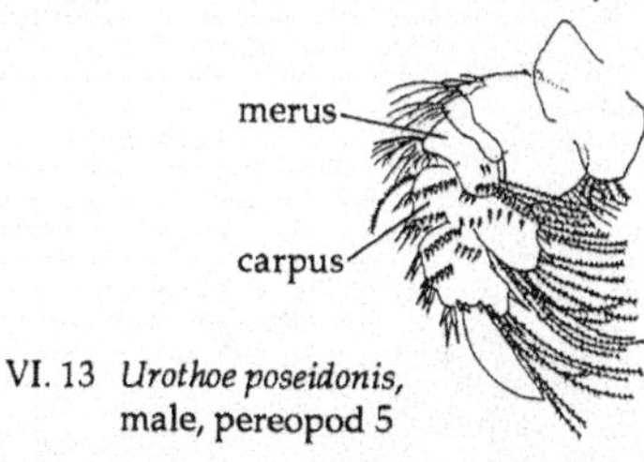

VI. 13 *Urothoe poseidonis*, male, pereopod 5

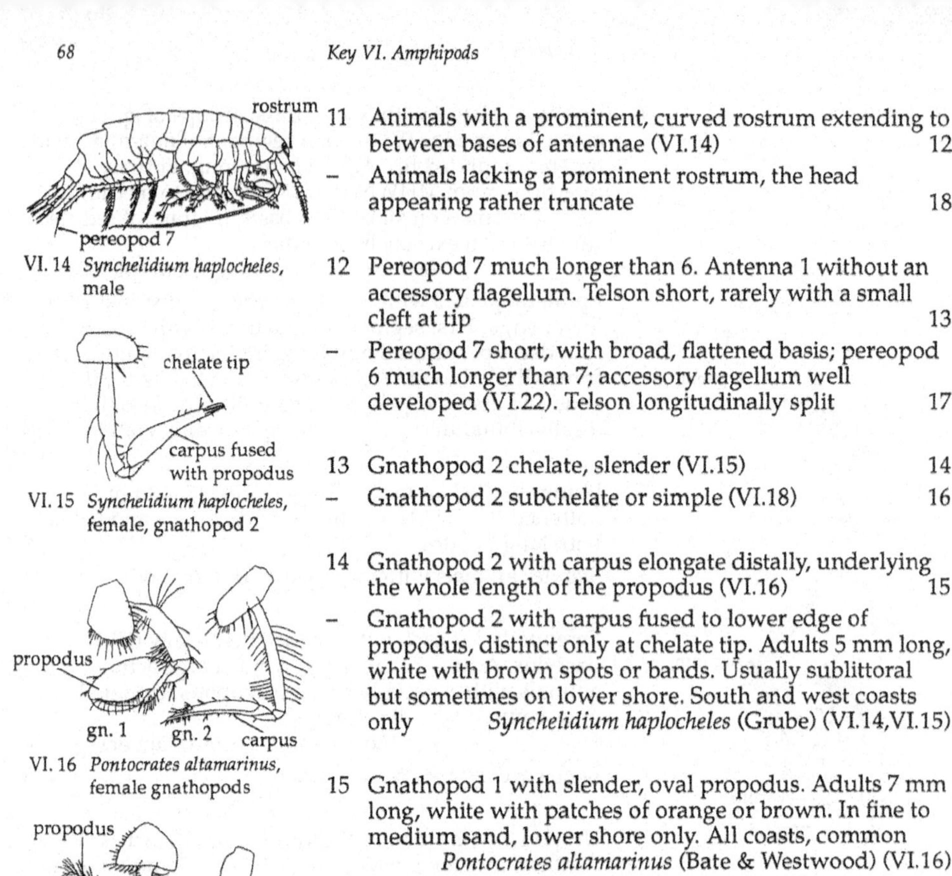

VI. 14 *Synchelidium haplocheles*, male

VI. 15 *Synchelidium haplocheles*, female, gnathopod 2

VI. 16 *Pontocrates altamarinus*, female gnathopods

VI. 17 *Pontocrates arenarius*, female gnathopods

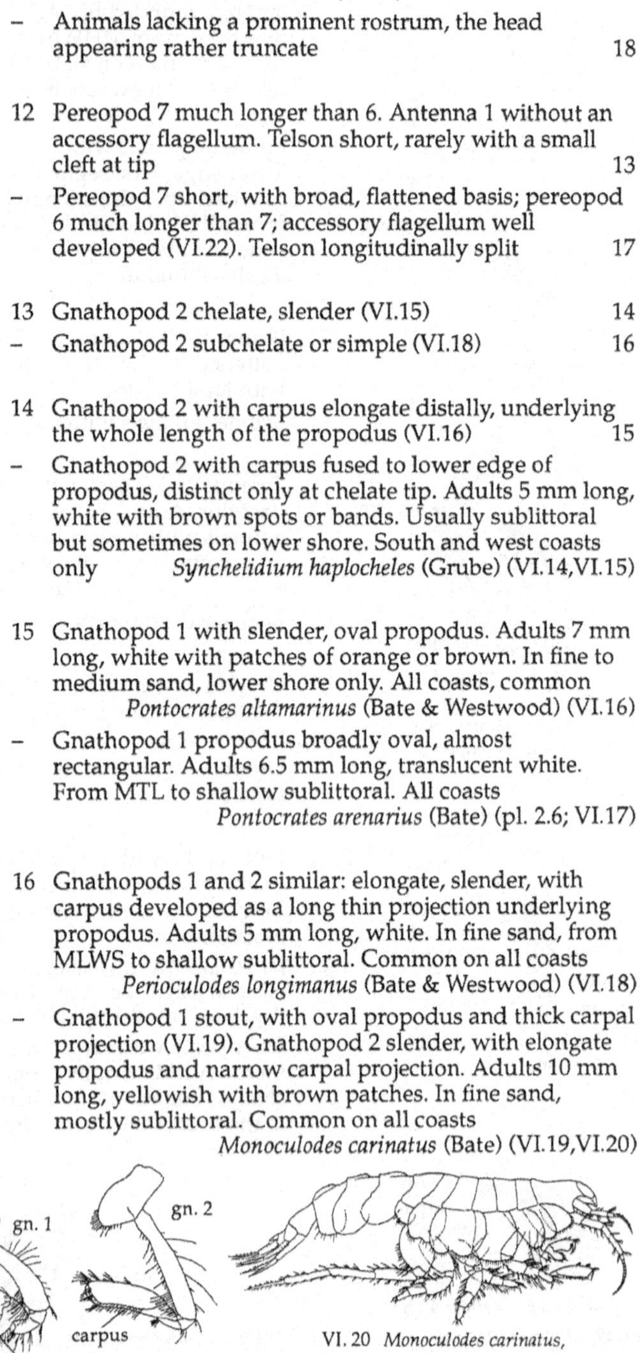

11 Animals with a prominent, curved rostrum extending to between bases of antennae (VI.14) — 12

– Animals lacking a prominent rostrum, the head appearing rather truncate — 18

12 Pereopod 7 much longer than 6. Antenna 1 without an accessory flagellum. Telson short, rarely with a small cleft at tip — 13

– Pereopod 7 short, with broad, flattened basis; pereopod 6 much longer than 7; accessory flagellum well developed (VI.22). Telson longitudinally split — 17

13 Gnathopod 2 chelate, slender (VI.15) — 14

– Gnathopod 2 subchelate or simple (VI.18) — 16

14 Gnathopod 2 with carpus elongate distally, underlying the whole length of the propodus (VI.16) — 15

– Gnathopod 2 with carpus fused to lower edge of propodus, distinct only at chelate tip. Adults 5 mm long, white with brown spots or bands. Usually sublittoral but sometimes on lower shore. South and west coasts only *Synchelidium haplocheles* (Grube) (VI.14,VI.15)

15 Gnathopod 1 with slender, oval propodus. Adults 7 mm long, white with patches of orange or brown. In fine to medium sand, lower shore only. All coasts, common *Pontocrates altamarinus* (Bate & Westwood) (VI.16)

– Gnathopod 1 propodus broadly oval, almost rectangular. Adults 6.5 mm long, translucent white. From MTL to shallow sublittoral. All coasts *Pontocrates arenarius* (Bate) (pl. 2.6; VI.17)

16 Gnathopods 1 and 2 similar: elongate, slender, with carpus developed as a long thin projection underlying propodus. Adults 5 mm long, white. In fine sand, from MLWS to shallow sublittoral. Common on all coasts *Perioculodes longimanus* (Bate & Westwood) (VI.18)

– Gnathopod 1 stout, with oval propodus and thick carpal projection (VI.19). Gnathopod 2 slender, with elongate propodus and narrow carpal projection. Adults 10 mm long, yellowish with brown patches. In fine sand, mostly sublittoral. Common on all coasts *Monoculodes carinatus* (Bate) (VI.19,VI.20)

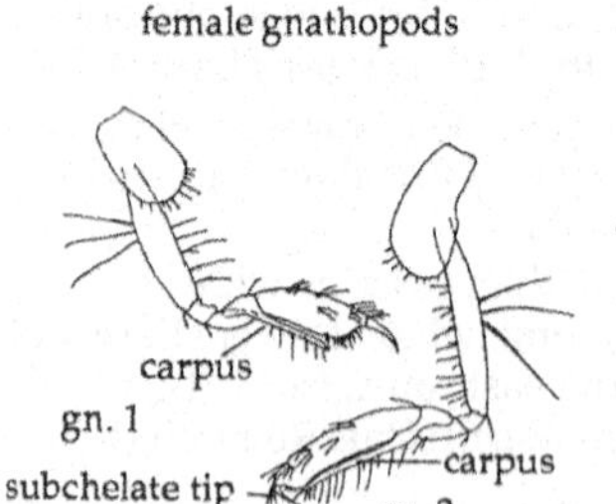

VI. 18 *Perioculodes longimanus* female gnathopods

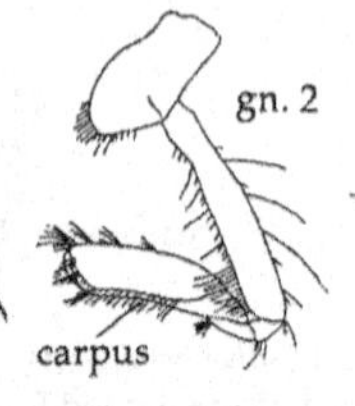

VI. 19 *Monoculodes carinatus*, male gnathopods

VI. 20 *Monoculodes carinatus*, female

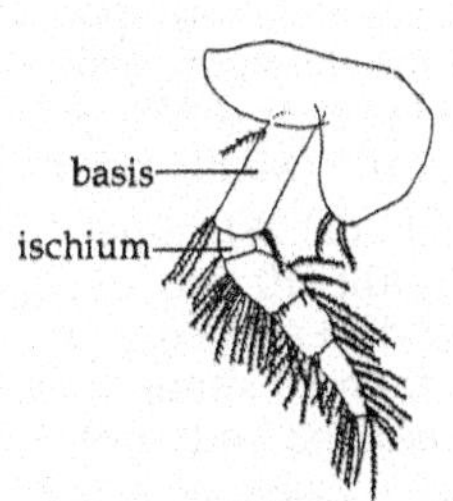

VI. 21 *Harpinia antennaria*, female, pereopod 5

17 Pereopod 5 with basis about as wide as ischium (VI.21). Pereopod 7 basis with eight or nine fine teeth on hind edge. Adults 5 mm long, white. On muddy or silty sand, mostly sublittoral but sometimes at ELWS. Common on all coasts *Harpinia antennaria* Meinert (VI.22)

– Pereopod 5 with basis more than twice width of ischium (VI.23). Rostrum very long, reaching to the end of antenna 1 peduncle. Adults 7 mm long, light brown to orange, with white flecks. In fine sand, lower shore and sublittoral. All British coasts *Phoxocephalus holbolli* (Krøyer) (VI.23)

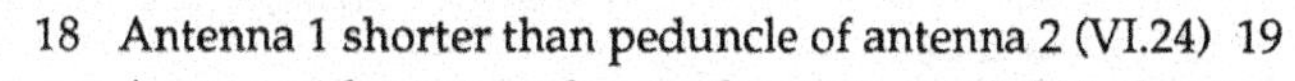

18 Antenna 1 shorter than peduncle of antenna 2 (VI.24) 19

– Antenna 1 longer or shorter than antenna 2, but always longer than antenna 2 peduncle 23

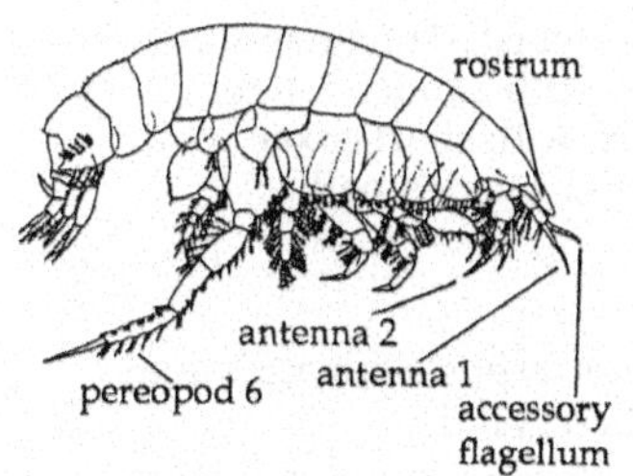

VI. 22 *Harpinia antennaria*, female

19 Uropod 3 with large, oval branches. Adults 5 mm long, with white, red and orange colouring. LWS and shallow sublittoral. All British coasts *Megaluropus agilis* Hoeck (VI.24)

– Uropod 3 with very short, cylindrical branches 20

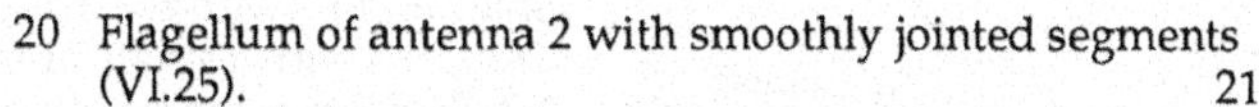

20 Flagellum of antenna 2 with smoothly jointed segments (VI.25). 21

– Flagellum of antenna 2 with small teeth at the joints between many segments (VI.27) 22

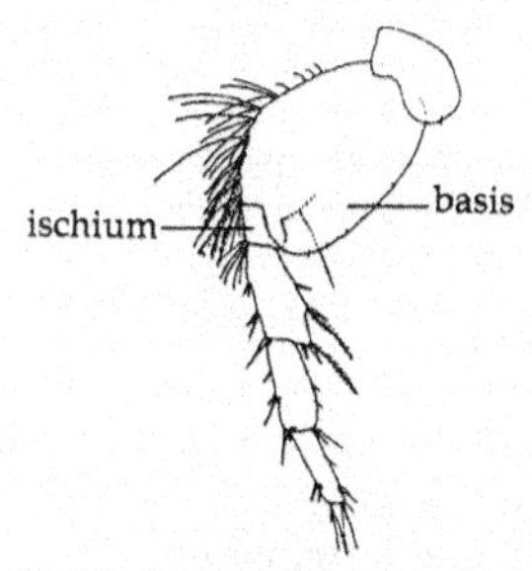

VI. 23 *Phoxocephalus holbolli*, female, pereopod 5

21 Pleopods with branches only about half length of stem (VI.25). Adults 18 mm long, reddish to greenish brown, banded with red. On shingle strands beneath drift algae. Common on all coasts *Orchestia gammarellus* (Pallas) (pl. 2.8)

– Pleopods with branches as long as or longer than stem (VI.26). Adults 17 mm long, brownish green. On shingle strands beneath drift algae. All coasts, but less common in North *Orchestia mediterranea* Da Costa (VI.26)

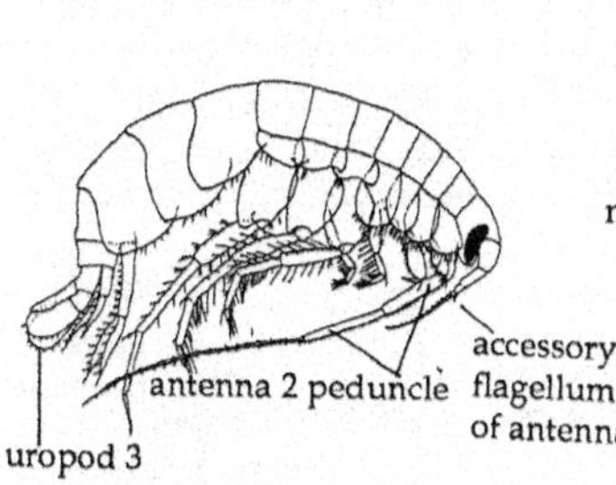

VI. 24 *Megaluropus agilis*, male

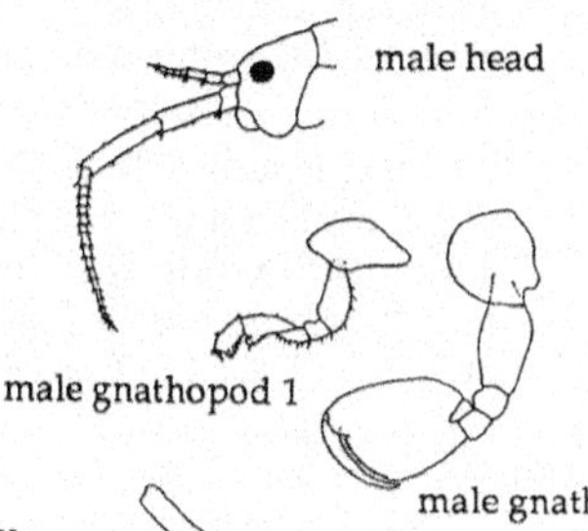

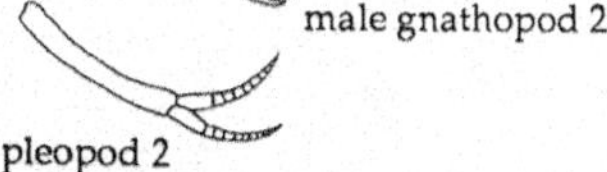

VI. 25 *Orchestia gammarellus*

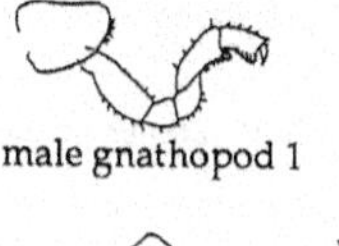

VI. 26 *Orchestia mediterranea*

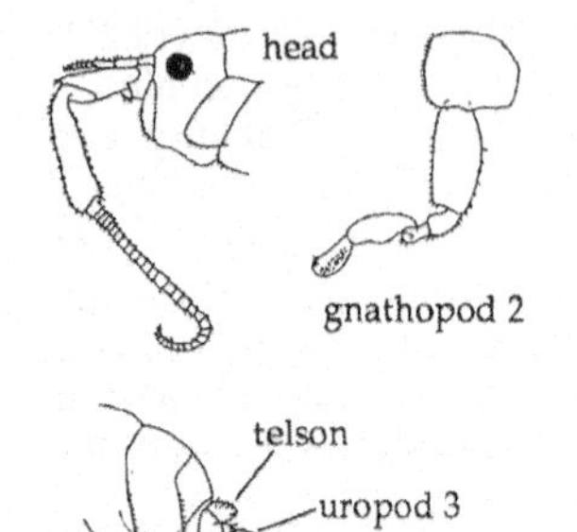

VI. 27 *Talitrus saltator*, male

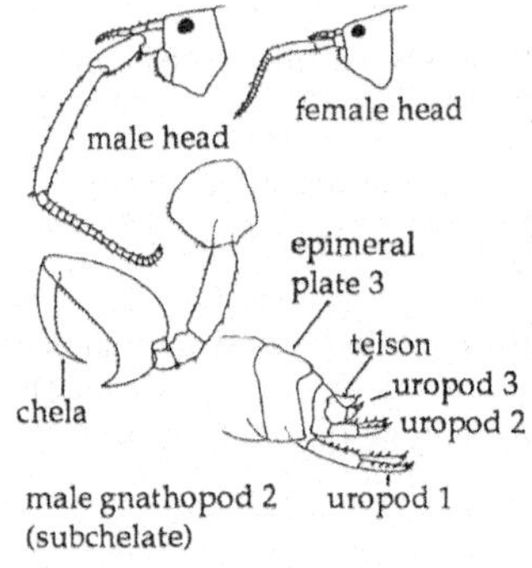

VI. 28 *Talorchestia deshayesii*

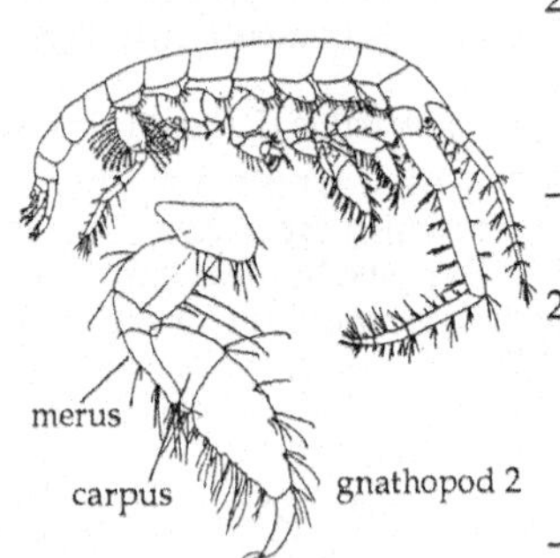

VI. 29 *Siphonoecetes kroyeranus*, male

22 Uropod 3 with long spine (VI.27). Male gnathopod 2 small, with tiny claw. Adults 20 mm long, brownish-grey. Along the strandline on sandy shores. Common on all coasts *Talitrus saltator* (Montagu) (pl. 2.4)

– Uropod 3 with several short spines (VI.28). Male gnathopod 2 large, subchelate. Adults 10 mm long, pale brown with dark stripes. Among drift algae on the strandline, on sand and shingle shores. Common on all coasts *Talorchestia deshayesii* (Audouin) (VI.28)

[*T. brito* Stebbing is larger, 15 mm long, yellowish, with uropod 3 branch almost as long as stem, and a rather square telson lacking a notch at tip. Rare; recorded from Aberdeen, Northumberland, The Wash, North Devon, Swansea.]

23 Body elongate and rather flattened. Antennae short and thick, often densely bristled, with 2 much longer than 1. Pereopods well spaced apart, with short coxae which scarcely touch (VI.29). Uropod 3 very small 24

– Not as described above 27

24 Gnathopod 2 with merus and carpus elongate, equal-sized and longitudinally fused (VI.31) 25

– Gnathopod 2 with normal merus and carpus (VI.29). Uropod 3 with a single branch, shorter than stem. Adults 5.5 mm long, with brown patch on top of head. Lower shore only, building tubes of sand grains. Recorded from east Scotland, southwest England and southwest Ireland *Siphonoecetes kroyeranus* Bate (VI.29)

25 Urosome segments fused, sides notched at insertion of uropods 1 and 2 (VI.30). Adults 5 mm long, brown. At LWS, burrowing in muddy sand. All British coasts *Corophium crassicorne* Bruzelius (VI.30)

– Urosome jointed and freely articulating 26

26 Uropod 1 peduncle with continuous row of short spines along outer edge (VI.31). Adults 8 mm long, white with brown patches. Intertidal, in sandy mud, often in estuaries. Common on all coasts *Corophium volutator* (Pallas) (pl. 2.5; VI.31)

– Uropod 1 peduncle with row of longer spines on outer edge, replaced towards base by fine bristles. Adults 6 mm, dull white with brown markings. Intertidal, building burrows in sand or muddy sand. South coasts only *Corophium arenarium* Crawford (VI.32)

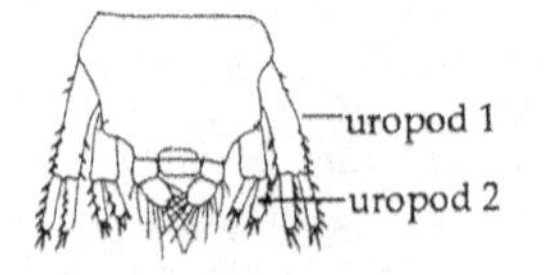

VI. 30 *Corophium crassicorne*, urosome viewed from above

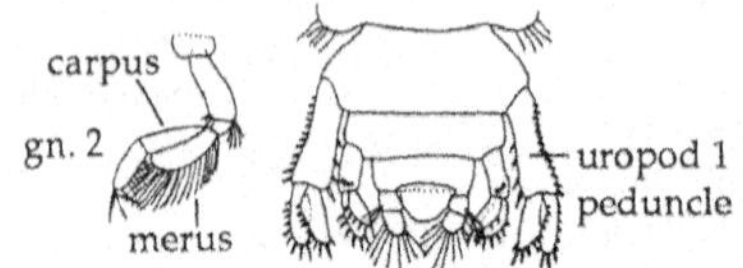

VI. 31 *Corophium volutator*, male urosome viewed from above

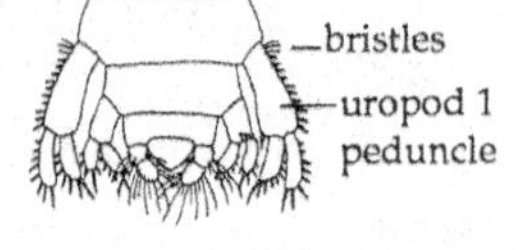

VI. 32 *Corophium arenarium*, urosome viewed from above

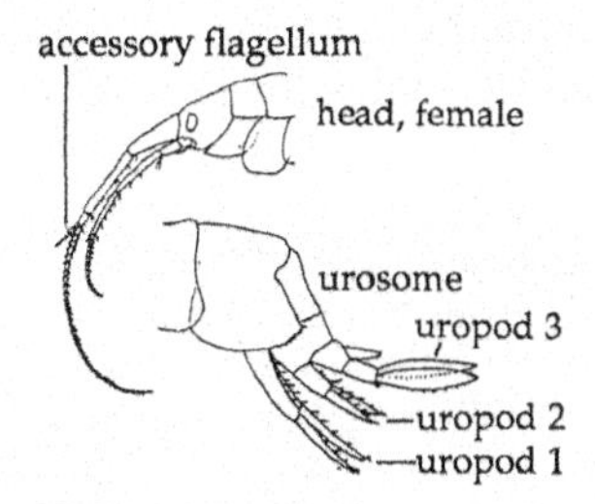

VI. 33 *Maera othonis*

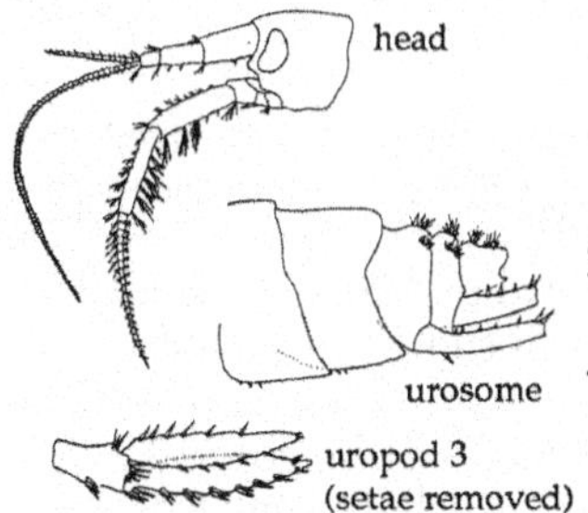

VI. 34 *Gammarus locusta*, male

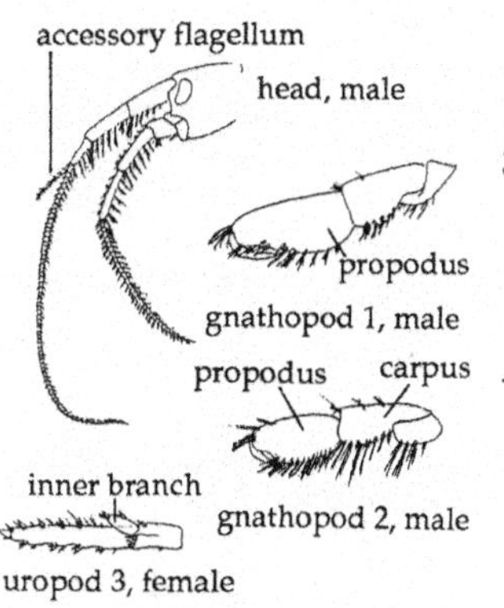

VI. 35 *Eulimnogammarus obtusatus*

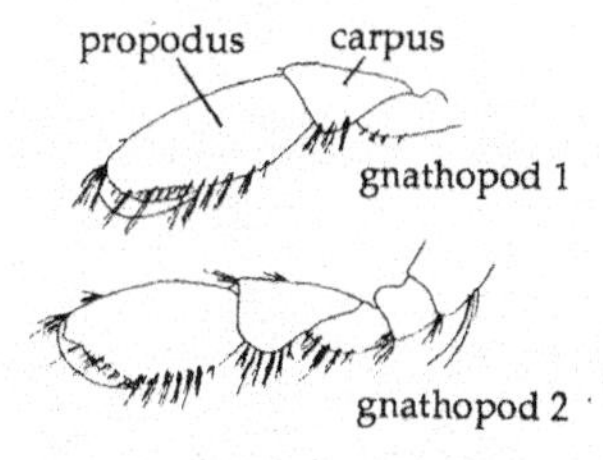

VI. 36 *Chaetogammarus marinus*, male, gnathopods

27 Antenna 1 with a well developed accessory flagellum (VI.33) 28

– Accessory flagellum absent or rudimentary 31

28 Head with small rostrum and large kidney-shaped eyes. Gnathopods 1 and 2 of similar size. Small groups of spines present on upper surface of urosome 29

– Head without rostrum, eyes small and round. Gnathopod 2 much larger than 1. Urosome without spines on upper surface. Adults 15 mm long, or larger, white with pale red markings. Lower shore only. All British coasts *Maera othonis* (Milne Edwards) (VI.33)

[*M. grossimana* (Montagu), which may occur beneath stones on shingly beaches, is distinguished by its gnathopod 2, the propodus of which is broadly expanded and appears almost rectangular in outline.]

29 Uropod 3 with inner branch less than one-third length of outer branch (VI.35) 30

– Uropod 3 with inner branch one-third length of outer, or longer. Adults up to 33 mm long. Intertidal and shallow sublittoral, often abundant on coarse grained shores *Gammarus locusta* (L.) (VI.34)

[Most species of *Gammarus* occur in brackish, estuarine or even fresh waters and are unlikely to occur on fully marine shores. However, river mouths, marshes or other sources of freshwater may permit some of these species to colonize adjacent beaches.]

30 Gnathopod 1 propodus longer than gnathopod 2 propodus. Gnathopod 2 with elongate carpus. Adults 20 mm long, brown or olive with purplish tints. Intertidal on shingly shores. All British coasts *Eulimnogammarus obtusatus* (Dahl) (VI.35)

– Gnathopod 1 propodus not longer than gnathopod 2 propodus (VI.36). Gnathopod 2 with short, triangular carpus. Adults 8–25 mm long, shades of green with red, yellow, brown or orange tints. Common on upper shore amongst gravel and shingle, not primarily sandy shore animals *Chaetogammarus* (VI.37)

[Three species occur commonly on British coasts. *C. marinus* Leach (VI.37) reaches 25 mm and is dark green; the pale green *C. pirloti* (Sexton and Spooner) reaches 15 mm; *C. stoerensis* (Reid) is pale green to bluish and only 8 mm long.]

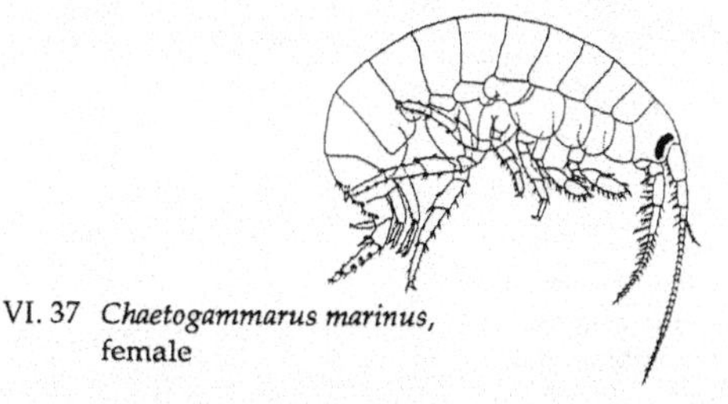

VI. 37 *Chaetogammarus marinus*, female

31 Gnathopods 1 and 2 both small and slender 32
– Either or both gnathopods 1 and 2 stout, with a robust propodus 36

32 Gnathopods simple, with slender propodus (VI.38). Pereopod 7 basis with a broad posterior lobe. Adults 12 mm long, translucent white, with brown or yellow spots. From MTL into shallow sublittoral, in sandy mud. Common on all coasts
Ampelisca brevicornis (Da Costa) (VI.38)
– Gnathopods with dactylus curved back, and thus distinctly subchelate (VI.39). Pereopod 7 basis without a posterior lobe 33

33 Head with short rostrum. Mandible with palp visible below peduncles of antennae (VI.40). Urosome segment 1 with two dorsal teeth (VI.41) 34
– Head with inconspicuous rostrum. Mandible without palp. Urosome segment 1 with only a single tooth 35

34 Distinct dorsal teeth on urosome only. Pereopod 3 dactylus shorter than propodus. Adults 8 mm long, translucent white with brown markings. Lower shore and sublittoral. Widespread and common on all British coasts
Atylus swammerdami (Milne Edwards) (VI.39–41)
– Dorsal teeth present on last segment of pereon and on all pleon and urosome segments. Pereopod 5 basis with pronounced hook on hind corner. Adults 8 mm long, translucent yellowish white, with orange patches. On fine sand and mud, lower shore only. All coasts
Atylus vedlomensis (Bate & Westwood) (VI.42)

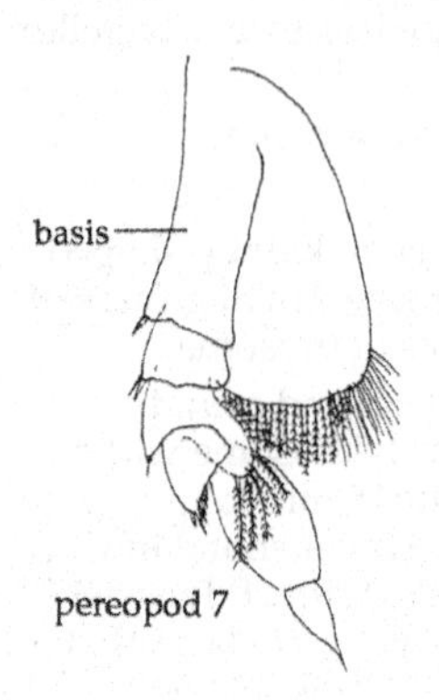

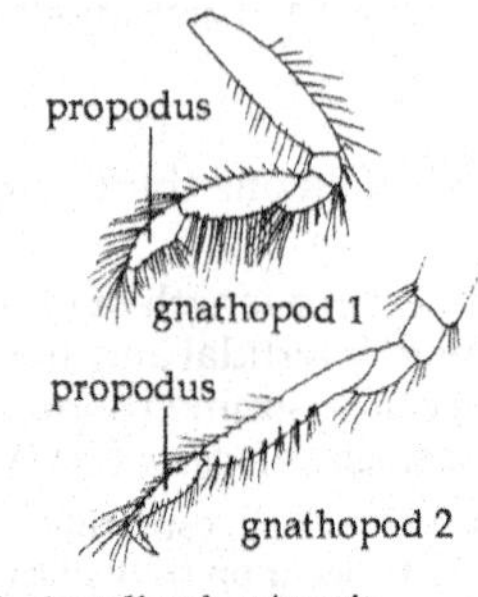

VI. 38 *Ampelisca brevicornis*

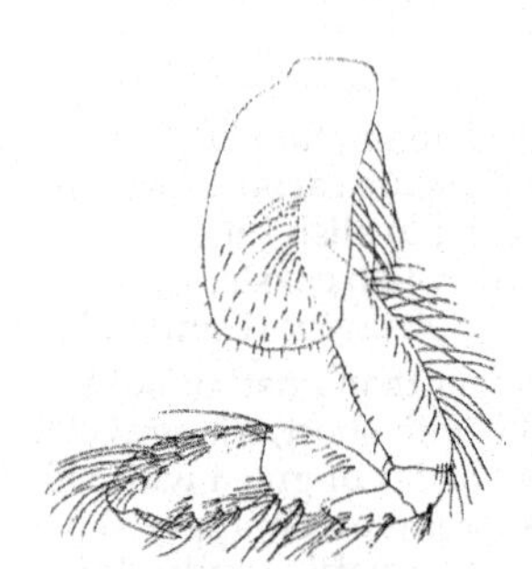

VI. 39 *Atylus swammerdami*, female, gnathopod 1

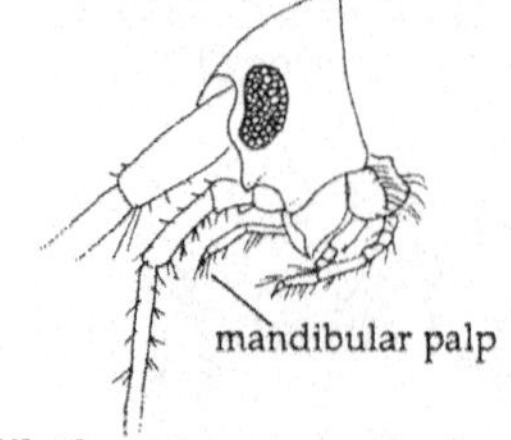

VI. 40 *Atylus swammerdami*, head, to show mandibular palp

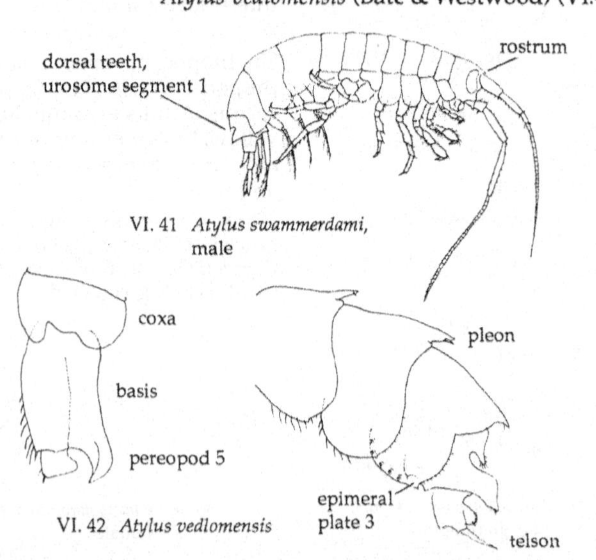

VI. 41 *Atylus swammerdami*, male

VI. 42 *Atylus vedlomensis*

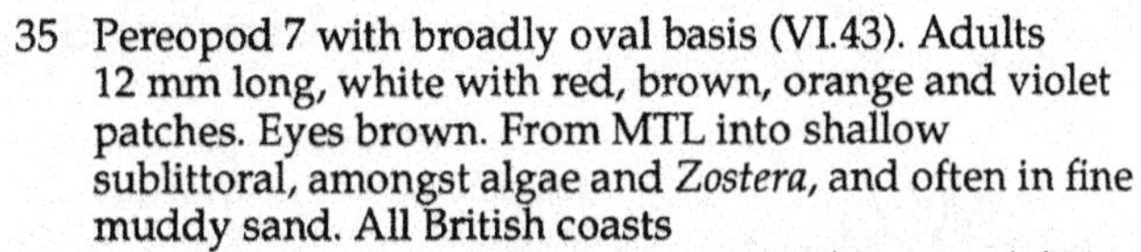

35 Pereopod 7 with broadly oval basis (VI.43). Adults 12 mm long, white with red, brown, orange and violet patches. Eyes brown. From MTL into shallow sublittoral, amongst algae and *Zostera*, and often in fine muddy sand. All British coasts
Dexamine spinosa (Montagu) (VI.44)

– Pereopod 7 basis slender, parallel-sided. Adults 5 mm long, yellowish green with white patches. Eyes red. Lower shore and shallow sublittoral, usually amongst algae. South and west coasts only *Dexamine thea* Boeck (VI.45)

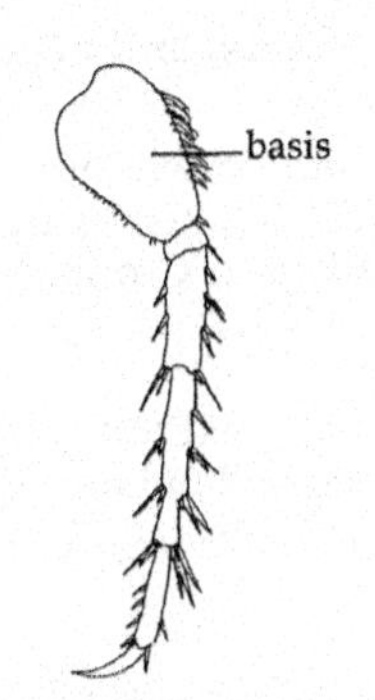

VI. 43 *Dexamine spinosa*, female, pereopod 7

36 Gnathopods 1 and 2 of similar size, each with carpus much shorter than propodus. Antenna 1 and 2 more or less equal, both rather slender. Adults 8 mm long. Lower shore, amongst algae. Common on all coasts
Calliopius laeviusculus (Krøyer) (VI.46)

– Gnathopod 2 larger than 1 37

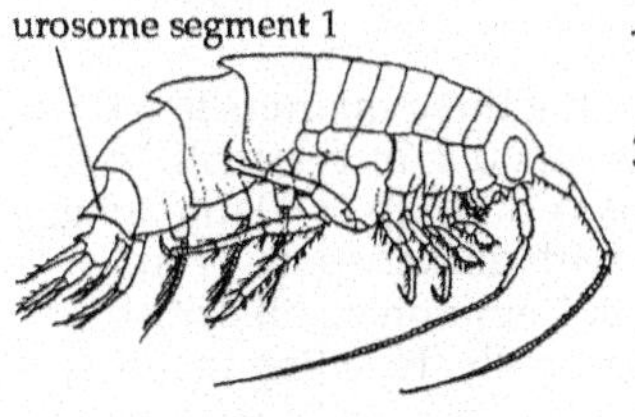

VI. 44 *Dexamine spinosa*, male

37 Gnathopod 2 with greatly enlarged propodus, and stout dactyl (VI.47). Carpus much smaller than propodus. Antenna 2 slightly longer than 1 but scarcely stouter. Adults 2 mm long, dark brown with lighter banding. Lower shore and sublittoral. Common on all coasts
Microprotopus maculatus Norman (VI.47)

– Gnathopod 2 stout, but not grossly enlarged; carpus about half length of propodus (VI.48). Antenna 2 slightly longer than 1 and about twice as thick. Lower shore and sublittoral, building tubes among stones and algae. Common on all coasts
Ampithoe rubricata (Montagu) (VI.49)

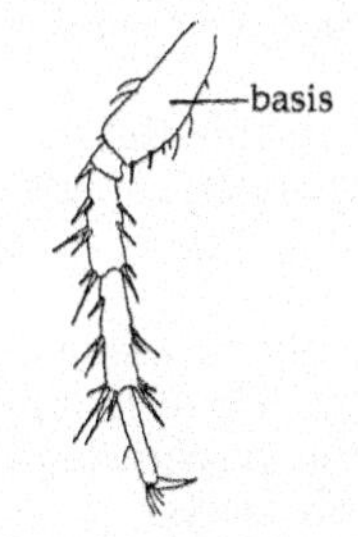

VI. 45 *Dexamine thea*, female, pereopod 7

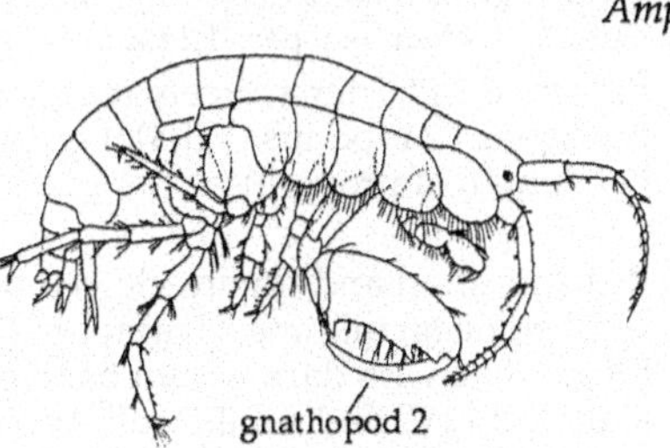

VI. 47 *Microprotopus maculatus*, male

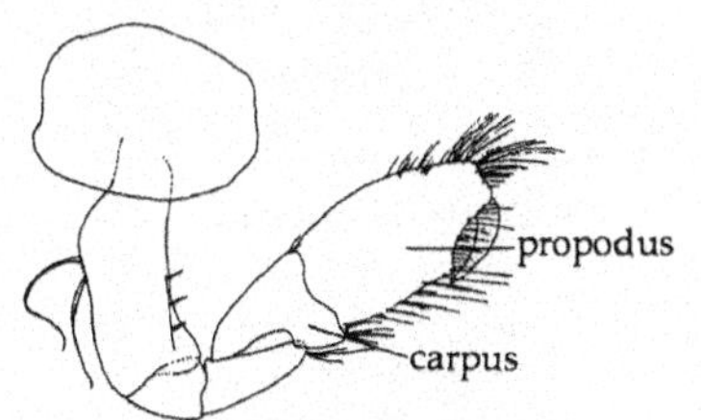

VI. 48 *Ampithoe rubricata*, male, gnathopod 2

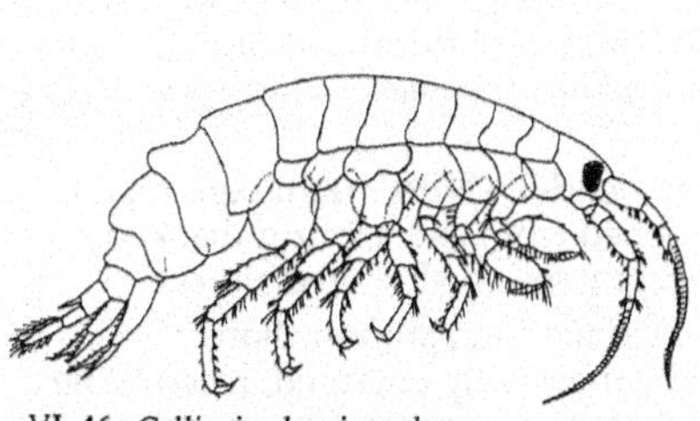

VI. 46 *Calliopius laeviusculus*, male

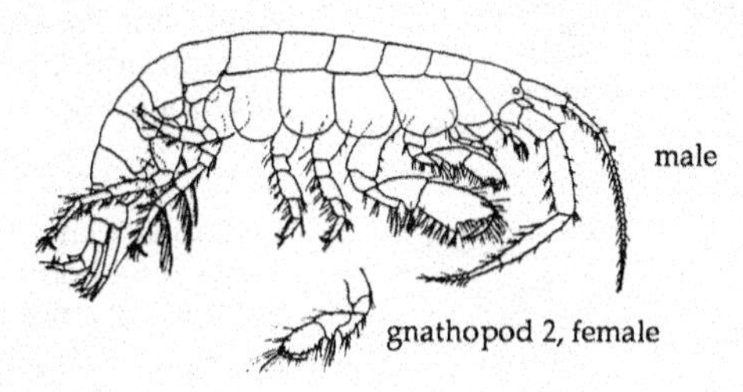

VI. 49 *Ampithoe rubricata*

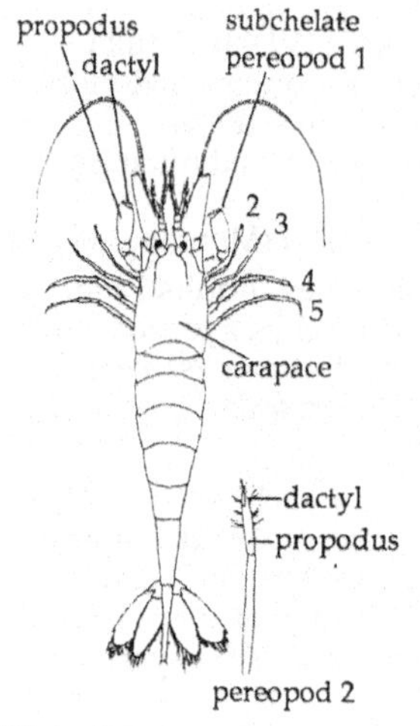

VII. 1 *Crangon crangon*

VII. 2 *Pontophilus fasciatus*, carapace viewed from above

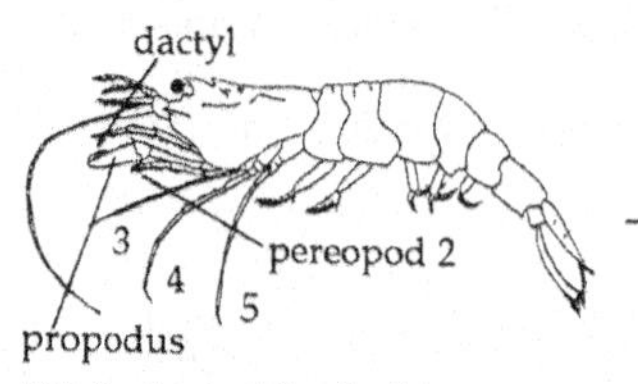

VII. 3 *Pontophilus fasciatus*

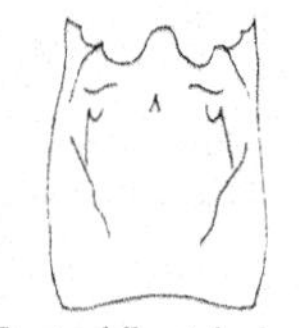
VII. 4 *Pontophilus trispinosus*, carapace viewed from above

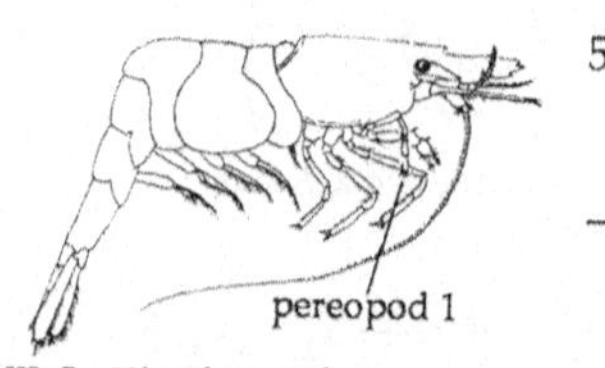

VII. 5 *Hippolyte varians*

Key VII. Decapods: Shrimps, Crabs and allies

All British species of shrimps and prawns are described and illustrated by Smaldon (1979). Crabs are covered by Ingle (1980, 1983) and Crothers & Crothers (1983).

1 Body shrimp-like, with carapace flattened from side to side, long abdomen and well-developed tail fan (VII.1) 2

– Body crab-like, with carapace flattened from top, and short abdomen turned up beneath (pl. 5) 10

2 First pereopod with dactyl folded over the end (subchelate) (VII.1) 3

– First pereopod pincer-like (chelate) slender or massive (VII.5–VII.10) 5

3 Pereopod 2 long, reaching along three-quarters length of propodus of pereopod 1; pereopod 2 dactyl small, less than half length of propodus (VII.1). Up to 90 mm long, grey to sandy brown, often with darker spots. From MTL into shallow sublittoral. Common on all British coasts; abundance varies seasonally; breeding involves migration into estuaries.
Brown shrimp *Crangon crangon* (L.) (VII.1)

[*C. allmani* Kinahan is a sublittoral species present on all British coasts but more common in the north. On northern shores occasional specimens may be found at LWS among *C. crangon*. *C. allmani* is distinguished by its last abdominal segment, which has a deep longitudinal groove on its upper surface, flanked by a parallel pair of ridges.]

– Pereopod 2 shorter, scarcely reaching basal quarter of pereopod 1 propodus (VII.3). Pereopod 2 dactyl at least half length of propodus 4

4 Rostrum short and square viewed from above; a single stout spine on midline of carapace (VII.2). Up to 19 mm long; white with dark brown band on pleon segments 4 and 6. At LWS and sublittoral. All British coasts
Pontophilus fasciatus (Risso) (VII.3)

– Rostrum short and rounded viewed from above; a single small spine on midline of carapace, and another on each side. Up to 27 mm long; yellowish brown, with dark mottling. At LWS and sublittoral. All British coasts
Pontophilus trispinosus Hailstone (VII.4)

5 Pereopod 1 small and slender, shorter than pereopods 3–5 (VII.5). Body flattened from sides, hump-backed. Swimming among seagrasses on sheltered shores 6

– Pereopod 1 (cheliped) longer and stouter than pereopods 3–5, usually massively enlarged, resembling a lobster claw. Body flattened from top. Burrowing in mud and muddy sand 7

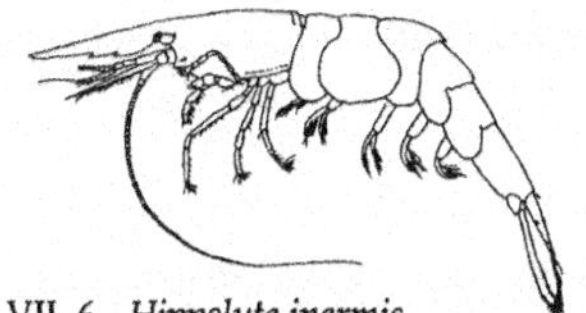
VII. 6 *Hippolyte inermis*

6 Rostrum shorter than carapace, with a sharp tooth at its base. Up to 32 mm long; red, brown or green, speckled with reddish brown. Lower shore and sublittoral. All British coasts *Hippolyte varians* Leach (VII.5)

– Rostrum longer than carapace, with smooth profile. Up to 42 mm long; unicolorous green, rarely brown or red. At LWS and in shallow sublittoral. South and west coasts only *Hippolyte inermis* Leach (VII.6)

[*H. longirostris* Czerniavsky occurs rarely at LWS on southwest coasts; it has a long rostrum with three or four teeth along its upper edge.]

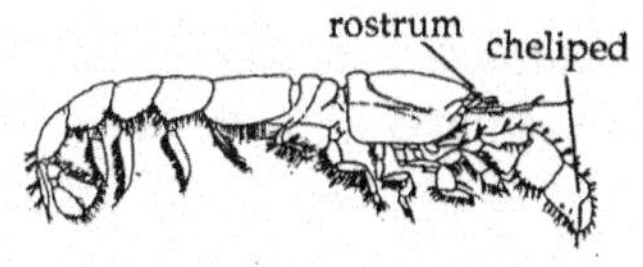

VII. 7 *Callianassa subterranea*

7 Left and right chelipeds unequal in size, one usually much larger than the other. Carapace with very short rostrum. Up to 40 mm long; pinkish white. Burrowing in muddy sand, at LWS and in shallow sublittoral. South coasts only, locally common *Callianassa subterranea* (Montagu) (VII.7)

[The larger *C. tyrrhena* (Petagna) is common on all British coasts but is entirely sublittoral. It is distinguished by its very short telson.]

– Left and right chelipeds equal-sized. Carapace with a large rostrum 8

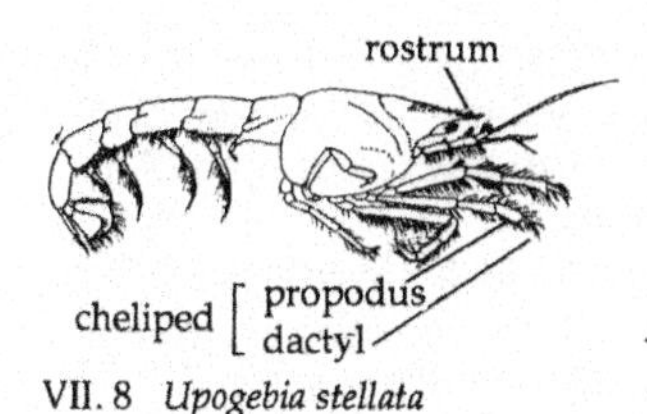

VII. 8 *Upogebia stellata*

8 Cheliped with slender propodus. Up to 50 mm long; yellowish white, often with orange spots. Burrowing in fine, muddy sand, ELWS and shallow sublittoral. All British coasts, common *Upogebia stellata* (Montagu) (VII.8)

– Cheliped with broad, massive propodus (VII.9,VII.10) 9

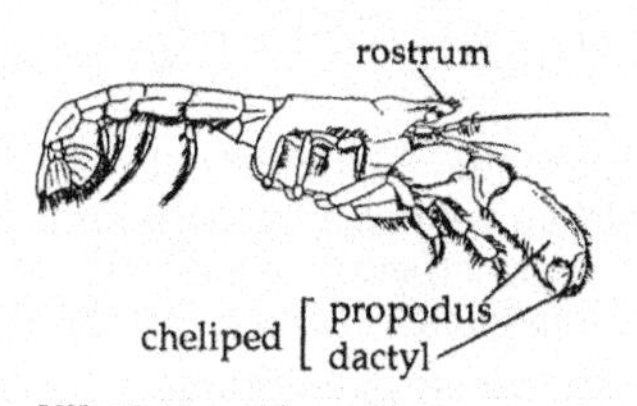

VII. 9 *Upogebia deltaura*

9 Cheliped with stout thumb, as long as dactyl (VII.9). Up to 100 mm or more; yellowish with white, green or reddish tints. In burrows of other animals, LWS and shallow sublittoral. All British coasts, common *Upogebia deltaura* (Leach) (VII.9)

– Cheliped with thumb shorter than dactyl. Up to 45 mm long. Burrowing in sand and muddy sand, ELWS and sublittoral. South coasts only, rare *Upogebia pusilla* (Petagna) (VII.10)

rostrum
cheliped [propodus dactyl]

VII. 10 *Upogebia pusilla*

10 Carapace wider than long. Antennae short 11

– Carapace longer than wide (VII.11). Antennae almost as long as carapace, stiff, with bristly inner edges. Male with very elongate chelipeds. Up to 39 mm carapace length; pale reddish to yellow. Lower shore and shallow sublittoral, burrowing in sand on all but the most exposed shores, with antennae maintaining respiratory channel to surface. Common on all British coasts Masked Crab *Corystes cassivelaunus* (Pennant) (pl. 5.1)

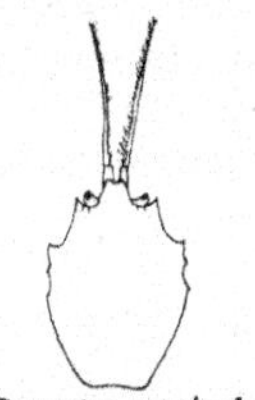
VII. 11 *Corystes cassivelaunus*, carapace

VII. 12 *Liocarcinus depurator*

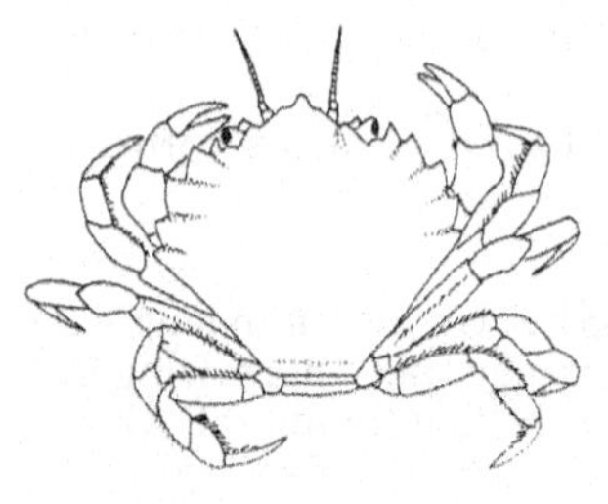
VII. 13 *Pirimela denticulata*

11 Dactyl of last pereopod thin and pointed — 12

– Dactyl of last pereopod flattened for swimming (VII.12) — 14

12 Carapace rounded, broadest at front, with dense fringe of stiff bristles. Carapace up to 20 mm long; pinkish brown. In fine sand, burrowing in topmost few cm. All British coasts, locally common
Thia scutellata (Fabricius) (pl. 5.4)

– Carapace with coarsely toothed front edge, without a bristly fringe — 13

13 One long tooth projecting from between eyes at front of carapace, with two short ones on each side. Chelipeds small, scarcely showing when the animal is still. Up to 18 mm carapace length; dark purplish, with red margins and white flecks, legs banded yellow and brown. LWS and sublittoral, burrowing in top few cm of sand. All British coasts, locally common
Pirimela denticulata (Montagu) (VII.13)

– Three short, equal-sized teeth between eyes at front of carapace. Chelipeds large and conspicuous. Up to 55 mm carapace length; colour variable, dark green to brown or reddish. Common, often abundant, on all rocky or gravelly shores and frequently vagrant on sand
Green Shore Crab *Carcinus maenas* (L.) (pl. 5.3)

14 Pereopod 5 with broadly rounded, flattened dactyl; pereopods 2–4 with cylindrical, pointed dactyls — 15

– Pereopods 2–5 all with flattened dactyls, those of pereopod 5 especially so. Carapace heart-shaped, as broad as long. Up to 20 mm carapace length; reddish brown with white patches. LWS and shallow sublittoral. All British coasts, perhaps more common in south
Portumnus latipes (Pennant) (pl. 5.2)

15 Carapace smooth. Up to 30 mm long, 35 mm broad; marbled light brown and reddish yellow. On sands and gravel, swimming freely. LWS and offshore. All British coasts, common *Liocarcinus marmoreus* (Leach) (pl. 5.6)

– Carapace with numerous short, transverse wrinkles bearing short rows of hairs. Up to 40 mm long, 51 mm broad; reddish brown, the flattened last pereopod with violet tip. On fine and mixed sands, swimming freely. LWS and sublittoral. All British coasts, common
Liocarcinus depurator (L.) (pl. 5.5)

Key VIII. Bivalves

DORSAL SIDE
projecting chondrophore
umbo
posterior adductor muscle scar
anterior adductor muscle scar
POSTERIOR
ANTERIOR
pallial sinus
pallial line
VENTRAL SIDE

VIII. 1 *Mya arenaria*, left valve

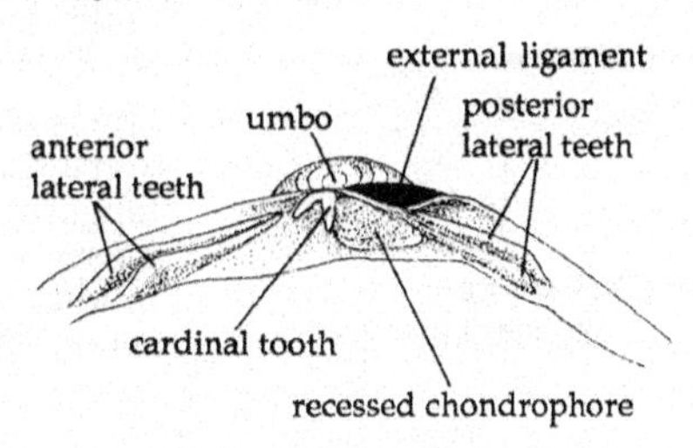

VIII. 2 *Mactra glauca*, hinge of right valve

With practice many bivalves may be identified by external features of living specimens, but for many others it is necessary to separate the two valves, remove the soft tissues, and examine the inner surfaces of the shells. Distinguishing between right and left shell valves is important, and there is no straightforward rule for doing this. The pallial sinus, when present, is always posterior, the umbones tend to point anteriorly, and the posterior adductor muscle scar is often larger than the anterior (VIII.1). It is important to distinguish between cardinal teeth and lateral teeth and to recognise the chondrophore (VIII.2), which supports the internal ligament in those species which possess it. It must be stressed that the following key is designed for the identification of live residents of sandy shores; dead shells of species from other habitats, which might be washed ashore infrequently, will not key out. All British bivalves may be identified using Tebble (1976).

VIII. 3 *Pharus legumen*, left valve

1 Razor shells: elongate, straight or slightly curved, with dorsal and ventral margins almost or quite parallel (pl. 4.1). Fast–moving, inhabiting deep, vertical burrows 2

– Not as described 7

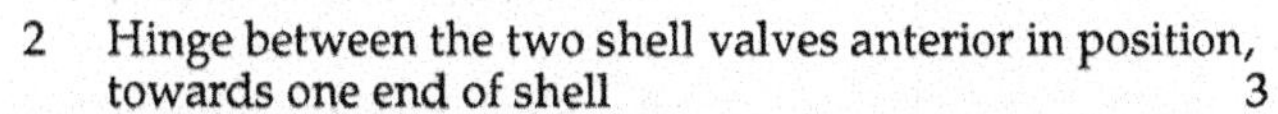

2 Hinge between the two shell valves anterior in position, towards one end of shell 3

– Hinge between shell valves more or less in the middle of the shell. Up to 140 mm long, rounded at both ends; white to light brown, with glossy, light green to yellow outer skin (periostracum). At LWS and into shallow sublittoral, southwest coasts of England, Wales and Ireland *Pharus legumen* (L.) (VIII.3)

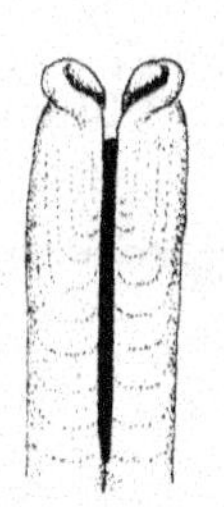

VIII. 4 *Solen marginatus*, anterior end of shell viewed from dorsal side

3 Each shell valve with a deep groove at the anterior end, close to the margin. Up to 130 mm long, straight, the two ends more or less squared off; yellowish white, with glossy light green to brown periostracum. At LWS and into shallow sublittoral, southern and western coasts only, not common *Solen marginatus* Montagu (VIII.4)

– Shell without such a groove 4

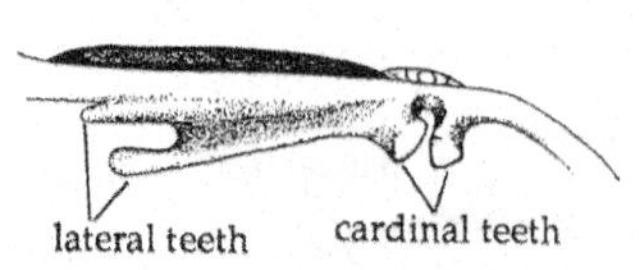

VIII. 5 *Ensis arcuatus*, hinge line of left valve

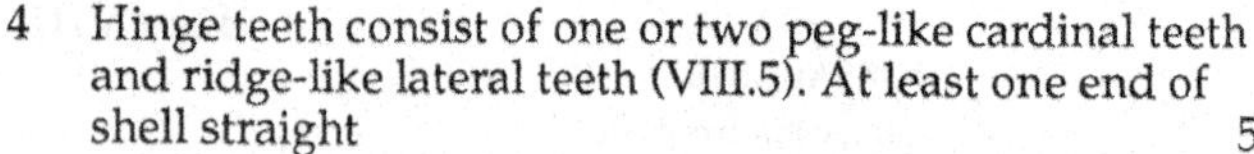

4 Hinge teeth consist of one or two peg-like cardinal teeth and ridge-like lateral teeth (VIII.5). At least one end of shell straight — 5

– Hinge teeth consist of one or two peg-like cardinal teeth and a similar sized lateral tooth (VIII.6). Up to 40 mm long, smoothly rounded at both ends; white, with glossy, green to yellowish brown periostracum. In muddy sands, mostly sublittoral, sometimes washed ashore. All British coasts — *Phaxas pellucidus* (Pennant) (VIII.6)

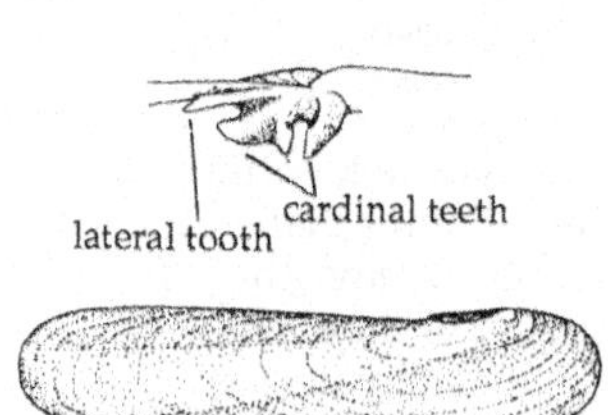

VIII. 6 *Phaxas pellucidus*

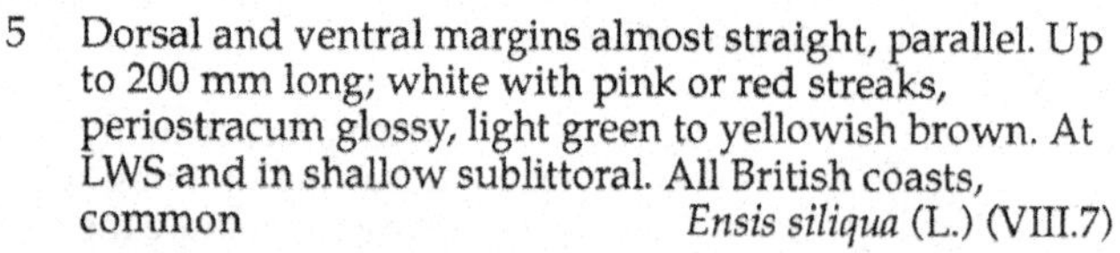

5 Dorsal and ventral margins almost straight, parallel. Up to 200 mm long; white with pink or red streaks, periostracum glossy, light green to yellowish brown. At LWS and in shallow sublittoral. All British coasts, common — *Ensis siliqua* (L.) (VIII.7)

– Either or both of dorsal and ventral margins curved — 6

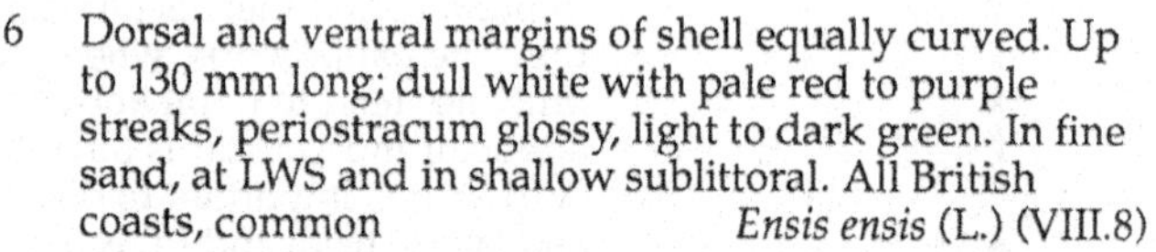

6 Dorsal and ventral margins of shell equally curved. Up to 130 mm long; dull white with pale red to purple streaks, periostracum glossy, light to dark green. In fine sand, at LWS and in shallow sublittoral. All British coasts, common — *Ensis ensis* (L.) (VIII.8)

– Dorsal margin of shell almost straight, ventral margin curved. Up to 150 mm long; white with red or purplish streaks, periostracum glossy, light green to yellowish brown. In sand and gravel, at LWS and in shallow sublittoral. All British coasts, common — *Ensis arcuatus* (Jeffreys) (pl. 4.1)

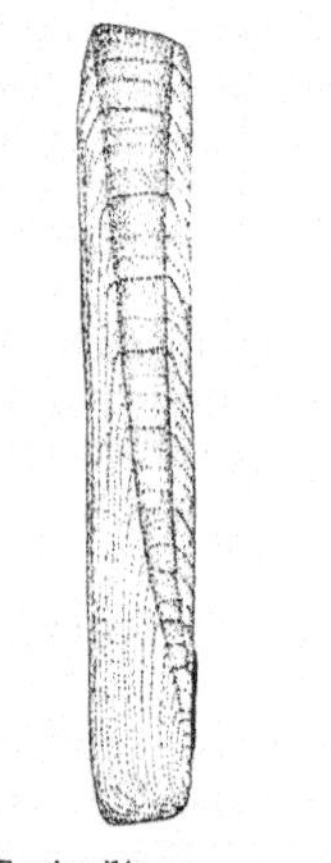

VIII. 7 *Ensis siliqua*, right valve

7 Cockles: plump, rounded shells with the two valves more or less equal; with well developed radiating ribs, often spiny (VIII.9–VIII.16). Shallow burrowers, the posterior part of the shell projecting at the surface — 8

– Not like this — 12

8 Shell thick, the inner surfaces smooth, except at the margin where the ribs on the outer surface are visible (VIII.9) — 9

– Shell thin, the ribs on the outer surface visible as grooves on the inner surface of the shell (VIII.10) — 11

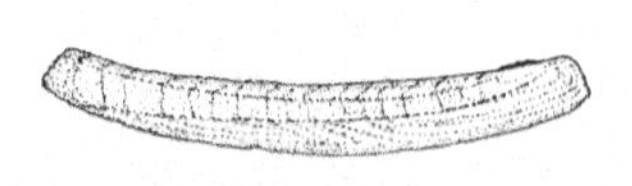

VIII. 8 *Ensis ensis*, right valve

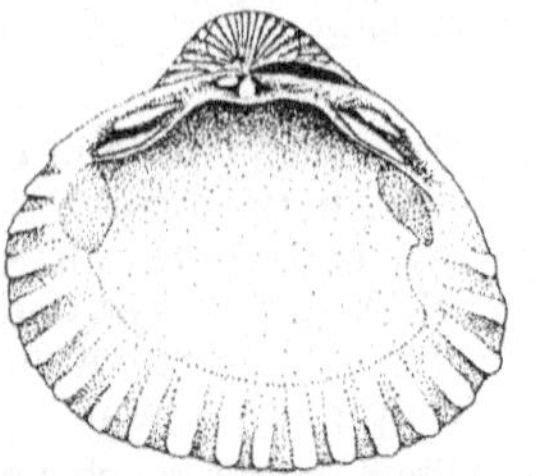

VIII. 9 *Cerastoderma edule*, inside of right valve

VIII. 10 *Acanthocardia echinata*, inside of right valve

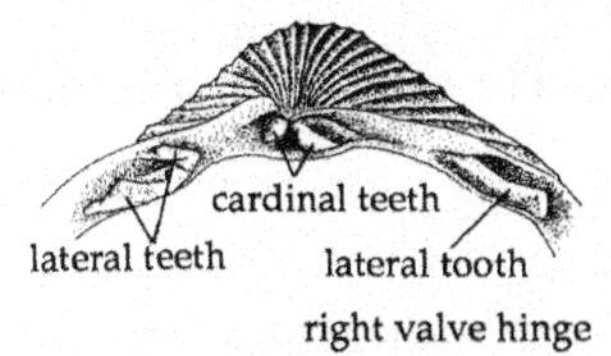

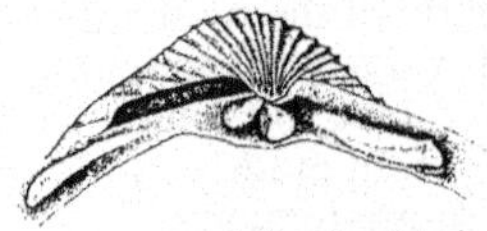

VIII. 11 *Acanthocardia tuberculata*

9 Right valve with one posterior lateral tooth; left valve with large anterior lateral tooth and smaller posterior lateral tooth (VIII.11) 10

– Right valve with two posterior lateral teeth (VIII.9); left valve with single, equally developed, anterior and posterior lateral teeth (VIII.12). Up to 50 mm long; cream to light yellow or pale brown, with thin yellowish or greenish brown periostracum. About 24 broad ribs, each with short, flattened, scale-like spines. In all sediments from muddy sand to fine gravel, MTL to ELWS. Common and abundant on all British coasts *Cerastoderma edule* (L.) (pl. 8.6)

[The Lagoon Cockle, C. *glaucum* (Poiret), is found in brackish water on southern coasts. It has a thinner, more asymmetric shell than C. *edule*, with the ribs visible on the inner surfaces.]

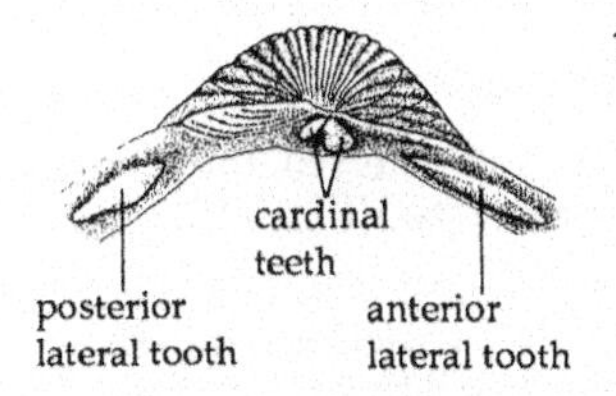

VIII. 12 *Cerastoderma edule*, left valve hinge

10 Large shell with 18 to 20 ribs, each bearing regularly spaced spines which close up towards the umbo to form a single spiny ridge. Up to 90 mm long; dull white to yellowish brown. In mixed muddy sands and fine gravel, ELWS and into shallow sublittoral. Uncommon on shore. South coasts only *Acanthocardia tuberculata* (L.) (VIII.11,VIII.13)

– Small shells with 20 or more ribs, occasionally bearing small tubercles but not spines *Parvicardium* species. (VIII.14)

[*P. scabrum* (Philippi) is up to 12 mm long with about 28 ribs; *P. exiguum* (Gmelin) is up to 13 mm long, with only 22 ribs. Both occur offshore on all British coasts but neither is found frequently between tidemarks.]

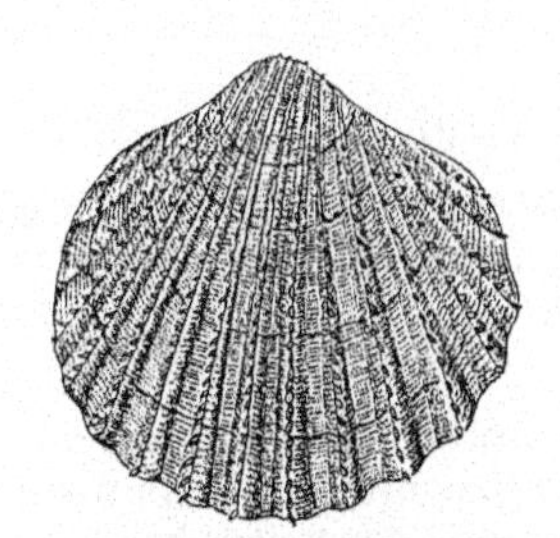

VIII. 13 *Acanthocardia tuberculata*, right valve

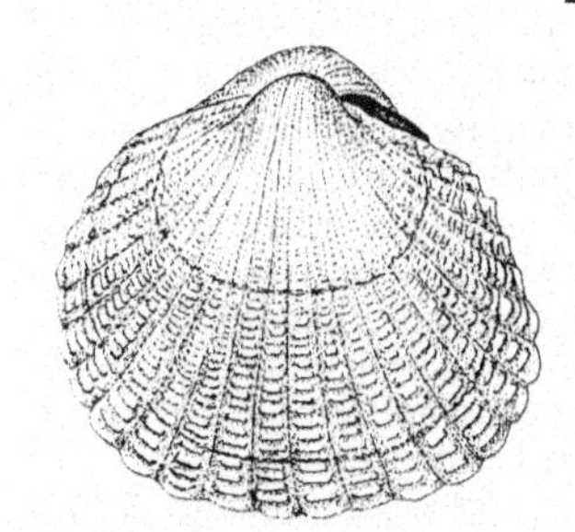

VIII. 14 *Parvicardium scabrum*, viewed from left side

11 Posterior margin of shell almost straight; about 20 prominent, sharply keeled ribs bearing regularly spaced spines, the longest and sharpest at the postero-ventral margin. Up to 100 mm long; dull white, yellow or light brown, with pinkish concentric bands. Sublittoral, rarely washed ashore. South and west coasts only *Acanthocardia aculeata* (L.) (VIII.15)

– Posterior margin of shell rounded; 18–23 prominent, keeled ribs with long, regularly spaced spines, the largest on the anterior portion of the shell. Up to 75 mm long; light to dark brown. In muddy sands and gravel, offshore only, occasionally washed ashore. All British coasts *Acanthocardia echinata* (L.) (VIII.10,VIII.16)

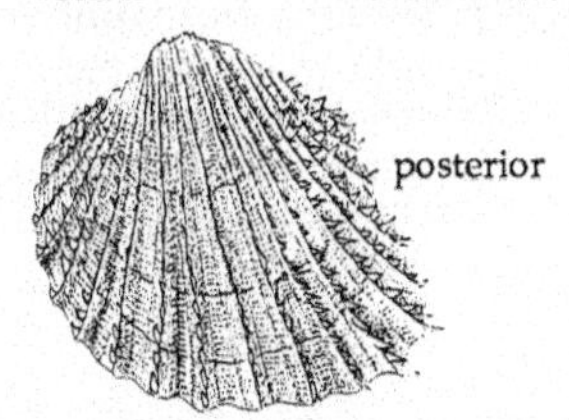

VIII. 15 *Acanthocardia aculeata*, left valve

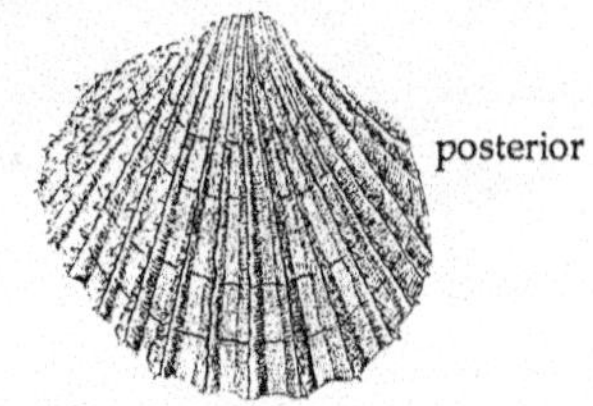

VIII. 16 *Acanthocardia echinata*, left valve

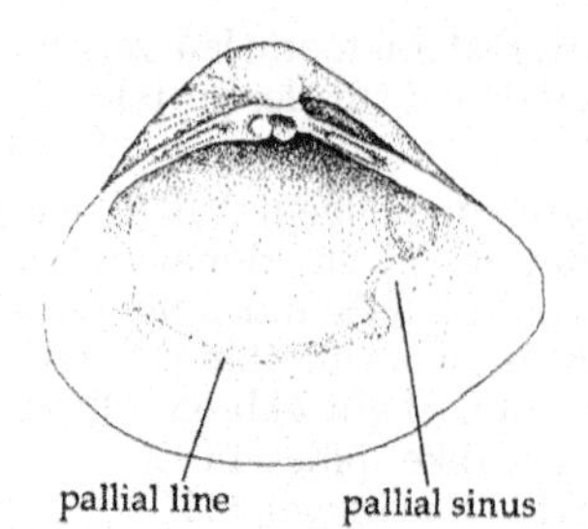

VIII. 17 *Corbula gibba*, inside of right valve

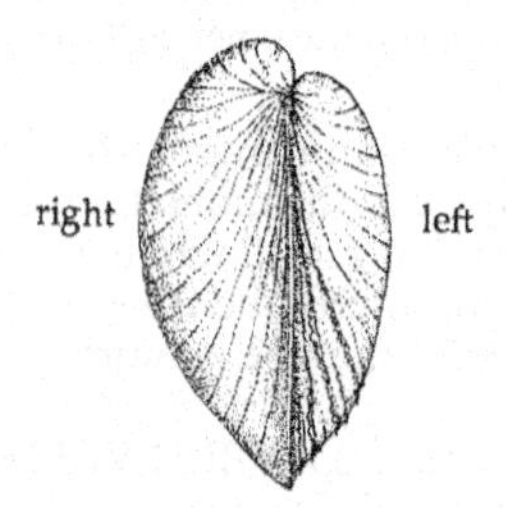

VIII. 18 *Corbula gibba*, viewed from anterior

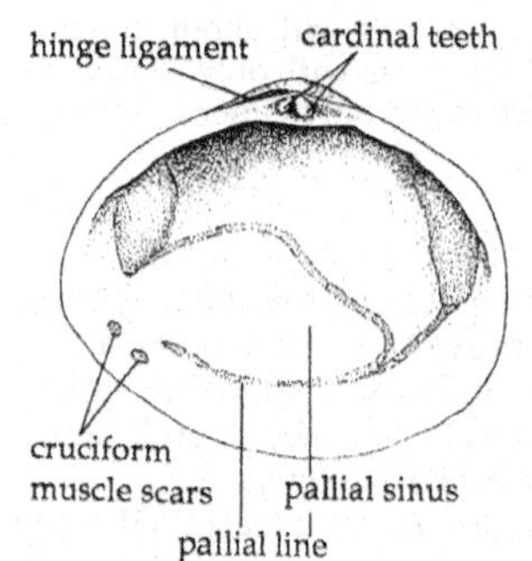

VIII. 19 *Macoma balthica*, inside view of left valve

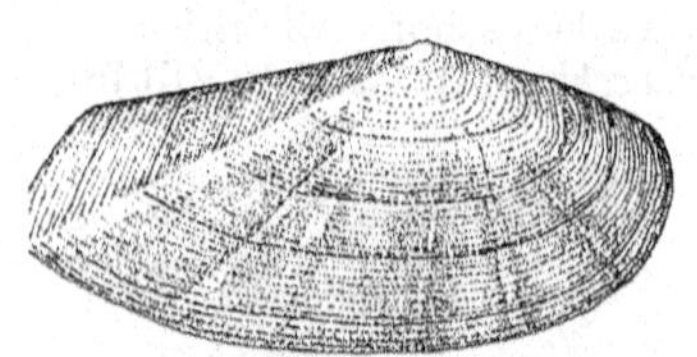

VIII. 20 *Gari fervensis*, right valve

12 Pallial line with a posterior sinus (VIII.17,VIII.19); sometimes small, and difficult to distinguish, but always present — 13

– Pallial line clearly without a sinus — 38

13 Shell nearly triangular; right valve more convex than left, enclosing it, the inner margin of the right valve covered only by a fringe of left valve periostracum. Pallial sinus small. Up to 15 mm long; off-white, with coarse, grey brown periostracum. In muddy sands and gravel, occasionally at ELWS, more frequent sublittorally. Common on all British coasts
Corbula gibba (Olivi) (VIII.17,VIII.18)

– Not as described; shell valves may be dissimilar but always close together along the margin — 14

14 Hinge ligament external (VIII.19) — 15

– Hinge ligament both internal and external, the internal part in a concave chondrophore (see VIII.1,VIII.2) — 31

15 Two cardinal teeth in each valve. Cruciform muscle scars present in many species beneath pallial line at posterior end of shell (VIII.19) — 16

– Three cardinal teeth present in each valve. No cruciform muscle scars — 23

16 Hinge line with cardinal teeth only, no lateral teeth — 17

– Hinge line with both cardinal teeth and lateral teeth, the latter conspicuous on the right valve, but developed only as thin ridges in the left — 18

17 Shell broadly oval, with the umbones more or less on the midline. Up to 25 mm long; white, yellow, pink or purple, one colour all over or with concentric bands of different colours. In mud and fine, muddy sands, intertidal only; most common in estuaries, less frequent on sea beaches. Common, and often abundant, on all British coasts *Macoma balthica* (L.) (pl. 4.8; VIII.19)

– Shell elongate-oval, more than twice as long as broad, with umbones in anterior half. Posterior end abruptly squared off. Each valve with a distinct ridge extending posteriorly from umbones. Up to 50 mm long; off-white to yellow, brown, red, pink or purple, usually in concentric bands, with a few radiating streaks of cream. In mixed medium to coarse sand and gravel, LWS and sublittoral. Common on most British coasts
Gari fervensis (Gmelin) (VIII.20)

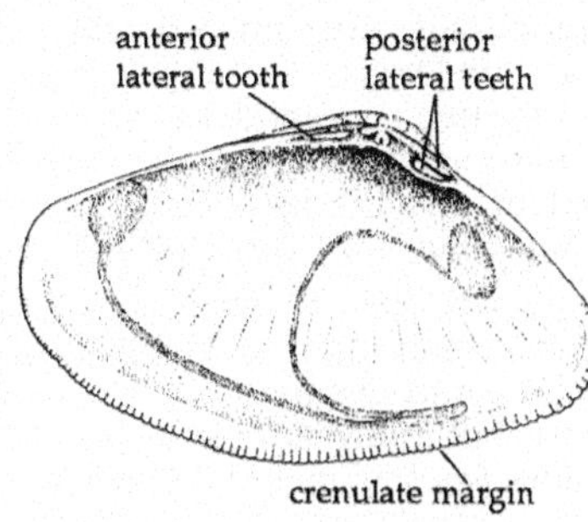

VIII. 21 *Donax vittatus*, inside of right valve

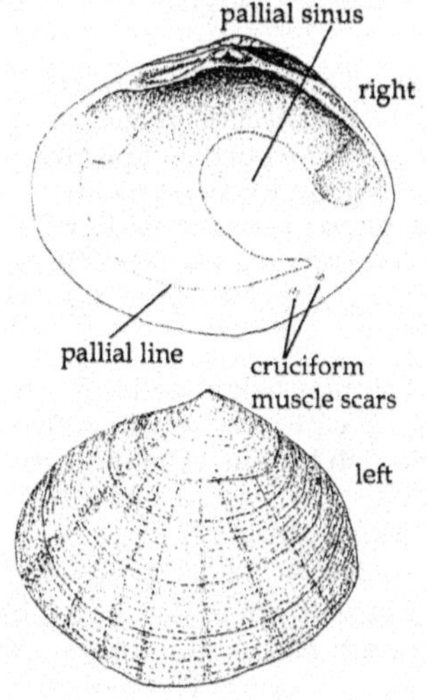

VIII. 22 *Arcopagia crassa*

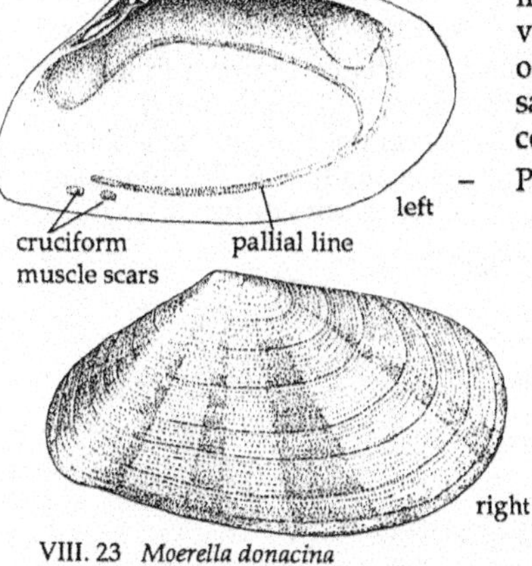

VIII. 23 *Moerella donacina*

18 Inner margin of shell smooth 19

– Inner margin of shell crenulate (finely wavy) (VIII.21). Broadly wedge-shaped, rounded anteriorly, tapered posteriorly, with the umbones in the posterior half. Up to 35 mm long; white to yellow or light brown, with a glossy, yellowish brown to olive periostracum. In fine to coarse sand, LWS to sublittoral. Common on all British coasts *Donax vittatus* (da Costa) (pl. 4.6)

[*D. variegatus* (Gmelin) occurs rarely on southern and southwestern coasts. It is distinguished principally by its much finer marginal shell crenulations, which feel smooth to the touch.]

19 Shell broadly oval, almost as deep as long. Lower edge of pallial sinus distinctly separate from the pallial line (VIII.22). Up to 60 mm long, with numerous fine concentric ridges; postero-ventral margin of shell with a shallow dent, concave on left, convex on right. Dull white to fawn, often pale orange to reddish around umbones, with a few reddish rays. In fine to medium sands, LWS and sublittoral. Southern and western coasts only *Arcopagia crassa* (Pennant) (VIII.22)

– Shell elongate-oval, at least 1.5 times long as deep. Lower edge of pallial sinus fused with pallial line (VIII.23) 20

20 Umbones situated in posterior half of shell. Small anterior and posterior lateral teeth in left valve. Elongate-oval, tapered posteriorly. Up to 25 mm long; yellowish white with bands and rays of cream, pink and red. In coarse sands and shell gravel, sublittoral but occasionally at ELWS. South and west coasts only *Moerella donacina* (L.) (VIII.23)

– Umbones close to midline of shell. No anterior lateral tooth in left valve 21

21 Posterior part of shell with concave dorsal and ventral margins - appearing hooked. Broadly oval, the two valves rather flat. Up to 45 mm long; fawn, yellow, orange or pink, often in variegated bands. In fine muddy sands, ELWS and sublittoral. Southern and western coasts, not common *Angulus squalidus* (Pulteney) (VIII.24)

– Posterior part of shell not hooked 22

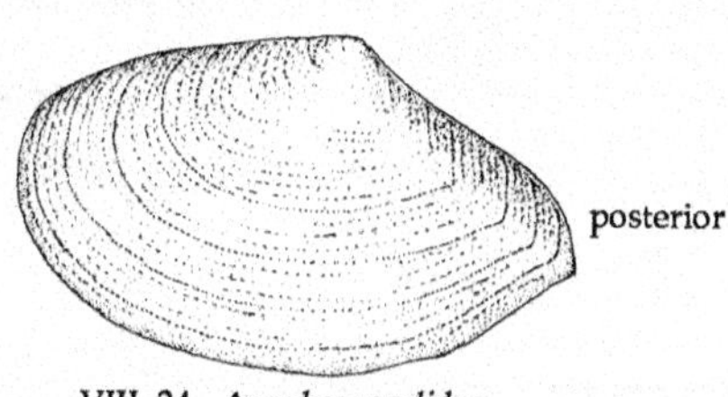

VIII. 24 *Angulus squalidus*, left valve

VIII. 25 *Fabulina fabula*, right valve

22 Shell nearly triangular, posterior half with convex dorsal and ventral margins. Up to 28 mm long; white to pink, yellow or orange, often in variegated concentric bands. In fine to medium sand, MTL to shallow sublittoral. Common, and usually abundant, on all British coasts
Angulus tenuis (da Costa) (pl. 4.7)

– Shell markedly tapered posteriorly. Up to 20 mm long; white, yellow or orange, often in concentric bands. In fine to medium sands, from LWS into shallow sublittoral. On all British coasts but rarely abundant
Fabulina fabula (Gmelin) (VIII.25)

23 Shell with bold concentric ridges (VIII.26–VIII.28) 24

– Shell with concentric grooves, or radiating ridges or striations, but without bold concentric ridges 27

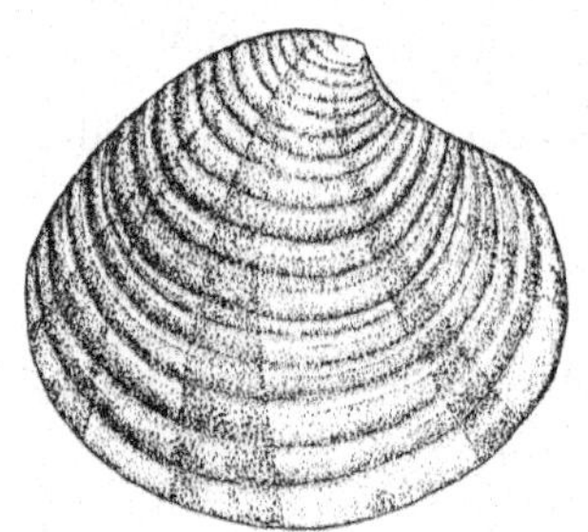
VIII. 26 *Clausinella fasciata*, right valve

24 Shell with up to 15 broad, flat ridges of variable width, giving a stepped appearance to its surface. Up to 25 mm long; white to yellow, brown, pink or purple, typically with rays or streaks of darker colour. In coarse sands and shell gravels, sublittoral, but occasionally at ELWS. All British coasts *Clausinella fasciata* (da Costa) (VIII.26)

– Shell with numerous, closely-spaced concentric ridges 25

25 Concentric ridges smooth, of similar height and thickness over whole shell. Dull white, cream or yellow, with three broad bands of reddish brown radiating from umbones. In fine to coarse sands, LWS and in shallow sublittoral. Common on all British coasts
Chamelea gallina (L.) (pl. 4.9)

– Concentric ridges coarse and prominent, often varying in height and width. Without three broad coloured rays 26

VIII. 27 *Venus verrucosa*, right valve

26 Concentric ridges intersecting with radiating ribs, particularly evident posteriorly where the ridges appear frilled. Up to 60 mm long; dull white to light brown. In fine to coarse sand and gravel, at LWS and shallow sublittoral. All British coasts *Venus verrucosa* (L.) (VIII.27)

– Concentric ridges prominent posteriorly but not frilled; no radiating ribs. Up to 50 mm long; dull white, tinted pinkish brown, with a few darker patches. In coarse sand and gravel, sublittoral but occasionally at ELWS. Common on all British coasts *Circomphalus casina* (L.) (VIII.28)

VIII. 28 *Circomphalus casina*, right valve

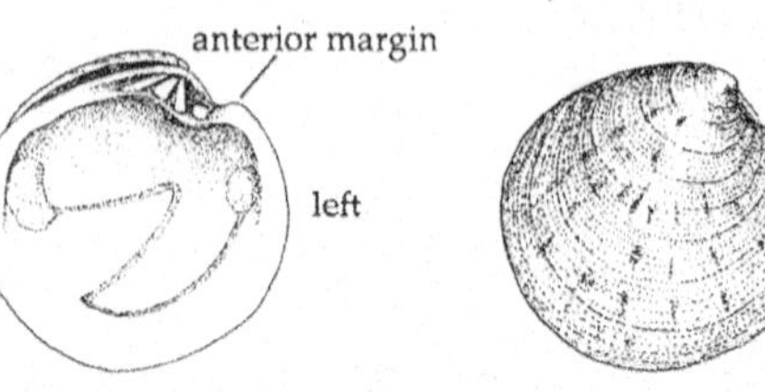

VIII. 29 *Dosinia exoleta*

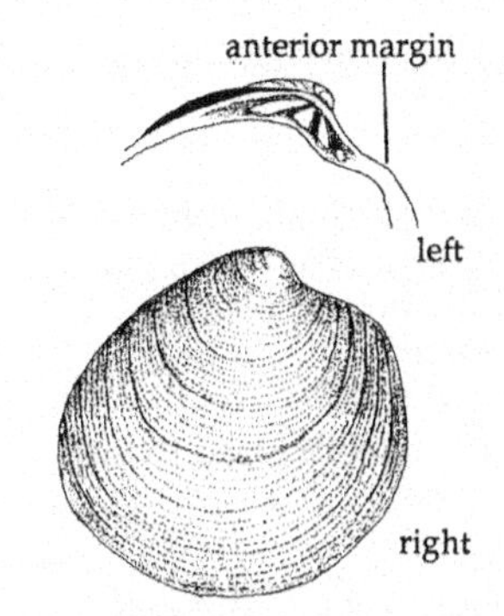

VIII. 30 *Dosinia lupinus*

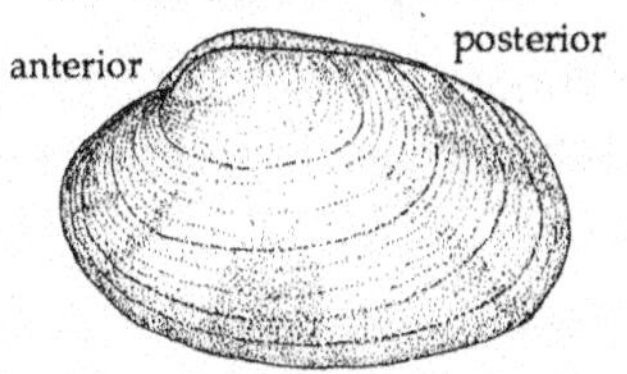

VIII. 31 *Paphia rhomboides*, viewed from left

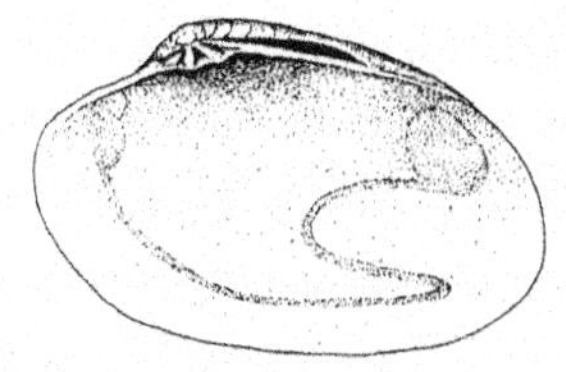

VIII. 32 *Paphia rhomboides*, inside of right valve

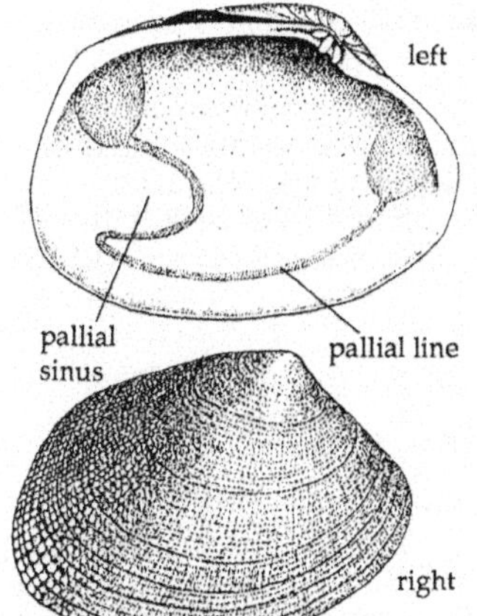

VIII. 33 *Tapes decussatus*

27 Shell almost circular, with the anterior margin arched (VIII.29, VIII.30) — 28

– Shell oval or rhomboidal, with the anterior margin sloping away from the umbones (VIII.31) — 29

28 Anterior margin strongly convex. Up to 60 mm long; white, yellow or brown, with darker, irregular rays, streaks or blotches of brown or pinkish brown. In coarse muddy sands and gravels, from LWS into sublittoral. All British coasts, common
Dosinia exoleta (L.) (VIII.29)

– Anterior margin gently convex. Up to 40 mm long; dull white to light brown, often yellow or pink dorsally. In mixed muddy sands and gravels. All British coasts
Dosinia lupinus (L.) (VIII.30)

29 Shell elongate-oval, with sculpture of concentric grooves only (VIII.31). Up to 60 mm long; cream, fawn or light reddish brown, with streaks, blotches or rays of chestnut brown, pinkish brown or purple. In coarse sands and gravel. Common on all British coasts
Paphia rhomboides (Pennant) (VIII.32)

– Shell with radiating ribs or lines, as well as concentric grooves — 30

30 Shell broadly oval or almost square; with fine concentric grooves and bold radiating lines, particularly pronounced posteriorly where the shell surface appears chequered. Pallial sinus not reaching midline of shell, its lower edge distinct from pallial line
Tapes decussatus (L.) (VIII.33)

– Shell elongate oval; with very fine concentric grooves and radiating striations, not pronounced posteriorly. Pallial sinus deep, extending beyond midline of shell, its lower edge partly fused with pallial line. Up to 50 mm long; cream to light brown, mottled with darker and lighter patches, bands and rays. In mixed fine to coarse sands, from LWS to shallow sublittoral. All British coasts *Venerupis pullastra* (Montagu) (VIII.34)

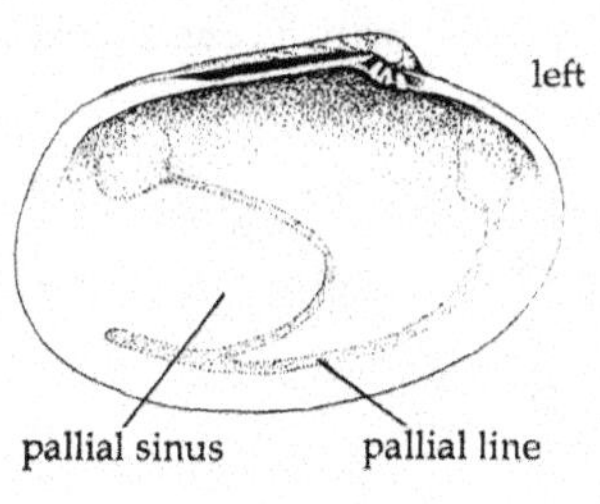

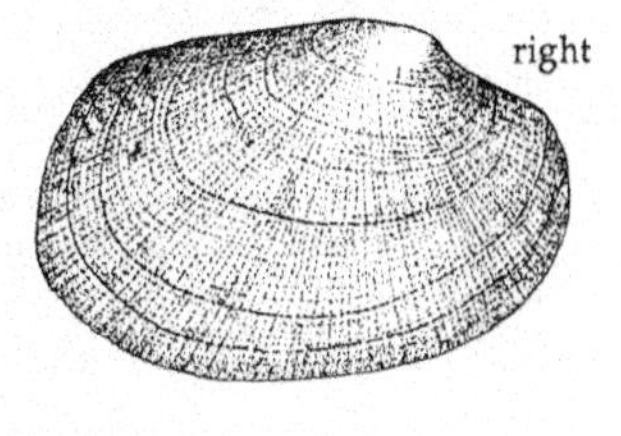

VIII. 34 *Venerupis pullastra*

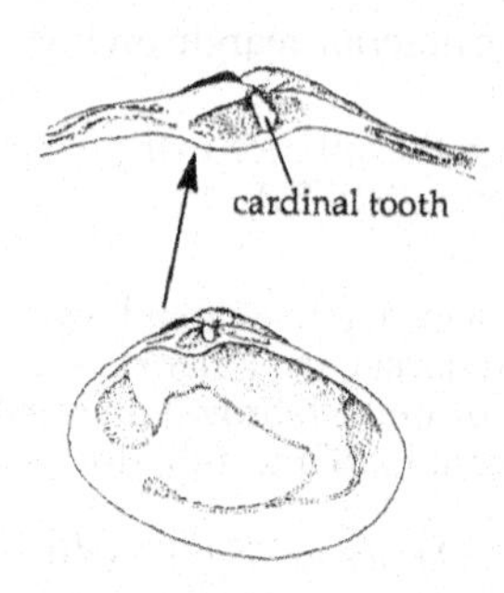

VIII. 35 *Abra alba*, inside of left valve

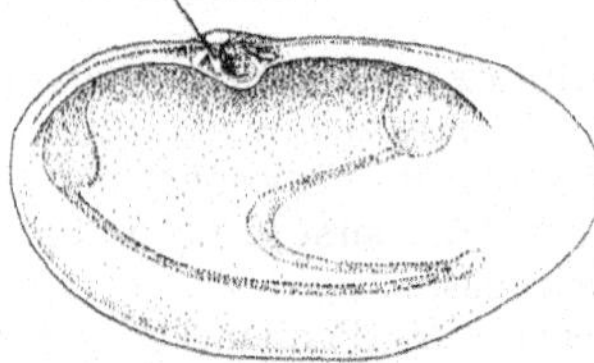

VIII. 36 *Lutraria lutraria*, inside of right valve

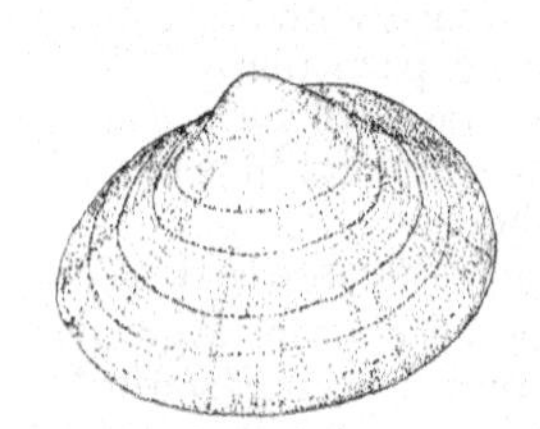

VIII. 37 *Mactra stultorum*, left valve

31 One or more cardinal teeth present on each valve — 32
– Hinge line without cardinal teeth — 36

32 Left valve with a single peg-like cardinal tooth, right valve with two small cardinal teeth. Shell oval, thin and brittle, up to 25 mm long; white. In mud and fine muddy sand, sublittoral but occasionally at ELWS. Common on all British coasts *Abra alba* (Wood) (VIII.35)
– Left valve with two stout cardinal teeth fused to form a solid ∧ shape (VIII.38) — 33

33 Shell elongate-oval, with umbones in anterior half, gaping widely at both ends. Up to 130 mm long; dull white to yellowish, sometimes tinted pink or purple. In mixed muddy sands, LWS to shallow sublittoral. All British coasts *Lutraria lutraria* (L.) (VIII.36)
– Shell nearly triangular, or oval; umbones on or close to midline of shell — 34

34 Shell thin and brittle. Lateral teeth with smooth interlocking surfaces. Up to 50 mm long; white to light purple, with light brown rays radiating from umbones. In fine to medium sand, lower shore and sublittoral. Common on all British coasts *Mactra stultorum* (L.) (VIII.37)

[*Mactra glauca* Born, which occurs on southwest coasts only, is larger. In its right valve, the anterior cardinal tooth is angled to the hinge line, rather than almost parallel to it (VIII.2).]

– Shell thick and solid. Lateral teeth with saw-toothed interlocking surfaces (VIII.38,VIII.39) — 35

35 Forked cardinal tooth of left valve small, extending only halfway down hinge plate. Up to 50 mm long; off-white to light brown, with greyish brown periostracum. In fine to medium sand, LWS into shallow sublittoral. All British coasts *Spisula solida* (L.) (VIII.38)
– Forked cardinal tooth of left valve larger, reaching almost to edge of hinge plate. Up to 30 mm long; dull white to cream, with greyish brown periostracum. In fine muddy or silty sand, LWS and shallow sublittoral. All British coasts *Spisula subtruncata* (da Costa) (VIII.39)

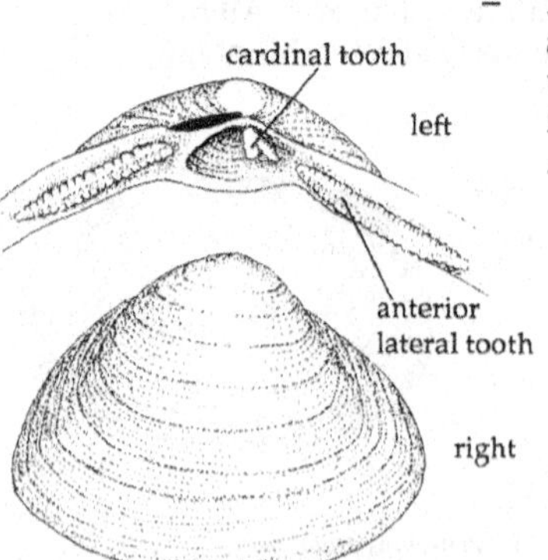

VIII. 38 *Spisula solida*

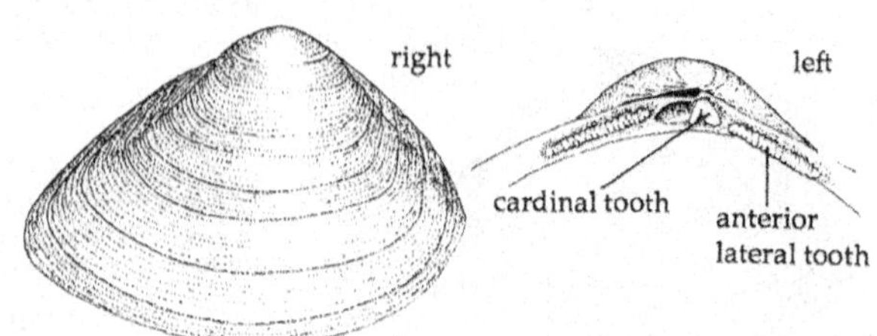

VIII. 39 *Spisula subtruncata*

36 Internal hinge ligament supported by a projecting, spoon-shaped chondrophore (VIII.1); prominent in left valve, concealed in right. Shell large, broadly oval, up to 150 mm long, widely gaping posteriorly where the thick, leathery siphons protrude. A deep static burrower in muddy sands; lower shore and shallow sublittoral, often abundant in estuaries. All British coasts, common *Mya arenaria* L. (pl. 8.7)

[*M. truncata* L. (pl. 8.8) has a similar distribution but is perhaps less common and rarely as abundant as *M. arenaria*. It has a shorter shell, with the posterior end abruptly truncated and the margin thus appearing straight.]

– Internal ligament situated in a concave chondrophore recessed into hinge plate (VIII.40). Shell not as described 37

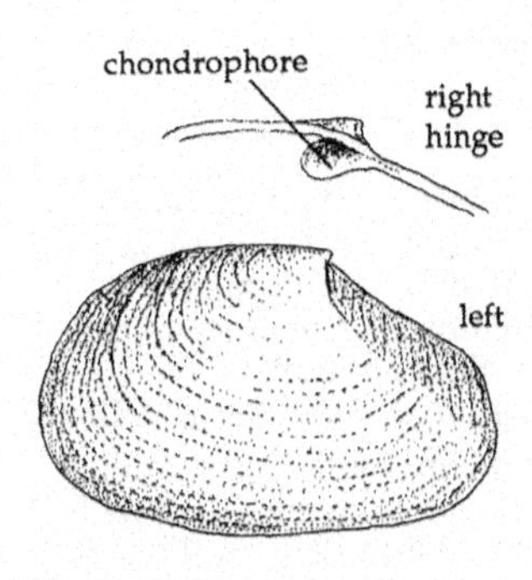

VIII. 40 *Cochlodesma praetenue*

37 Each valve with a short but distinct crack in its calcification just posterior to the umbo. Shell oval, up to 40 mm long; dull white with a light yellowish brown periostracum. In silty sand, from LWS into sublittoral. All British coasts *Cochlodesma praetenue* (Pulteney) (VIII.40)

– Valves without such a crack. Shell oblong, 30–40 mm long; dull white with light yellowish brown periostracum. In mixed fine to coarse sands. All British coasts *Thracia* (VIII.41)

[Two species occur commonly, *T. phaseolina* (Lamarck) and *T. villosiuscula* (Macgillivray), but are difficult to tell apart. The latter has a finely granular shell surface, with the granulations visible with a x10 hand lens.]

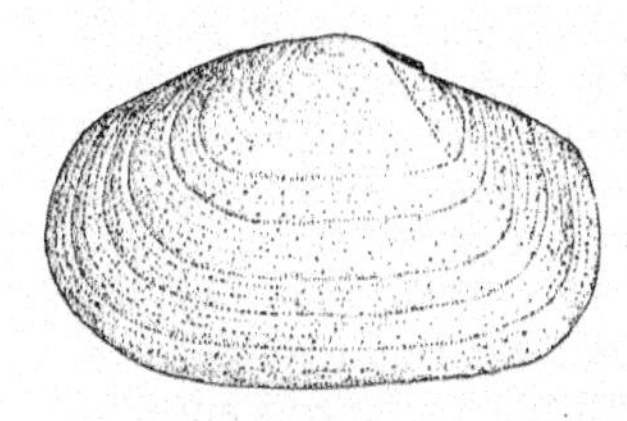

VIII. 41 *Thracia phaseolina*, left valve

38 Three cardinal teeth in each valve, hinge ligament external. Shell oval to circular, with prominent umbones, up to 120 mm long; dull white with a thick, glossy periostracum, deep greenish brown in juveniles, black in the largest specimens. In fine to coarse muddy sand, offshore but occasionally at ELWS. All British coasts *Arctica islandica* (L.) (VIII.42)

– Three cardinal teeth in one valve only, or both valves with less than three 39

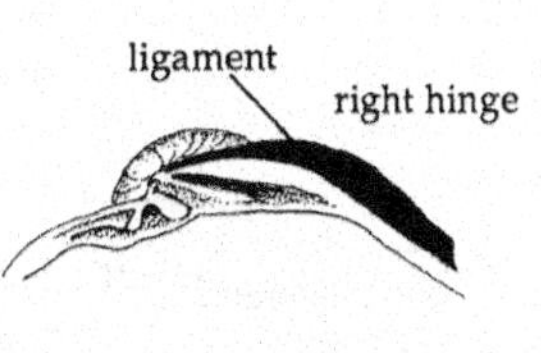

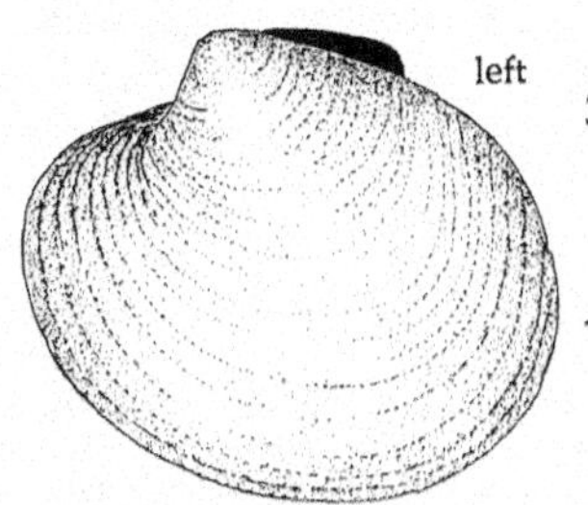

VIII. 42 *Arctica islandica*

39 Shell thin, flat, almost square, its surface densely pitted. Up to 12 mm long; white. In silty sands from ELWS into shallow sublittoral, in burrows of the burrowing prawn *Upogebia* *Lepton squamosum* (Montagu) (VIII.43)

– Shell convex, oval, surface not pitted. 40

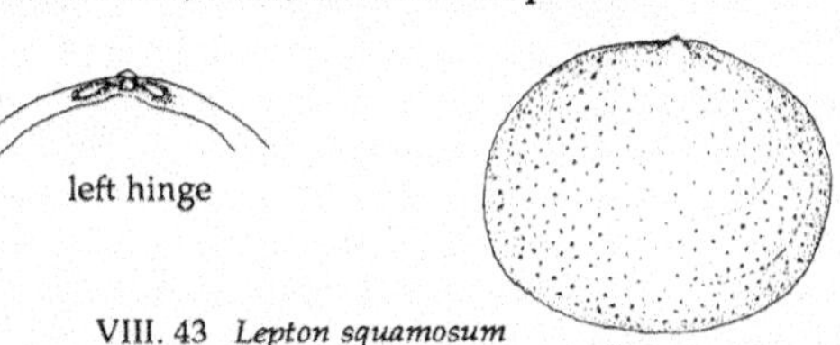

VIII. 43 *Lepton squamosum*

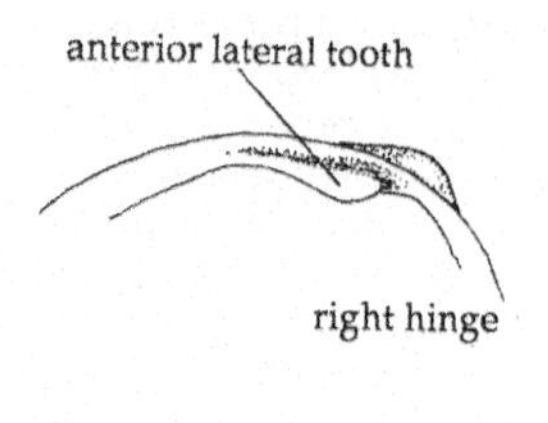

40 Hinge with anterior lateral teeth in one or both valves, but no posterior lateral teeth — 41

– Hinge with one anterior lateral tooth and one posterior lateral tooth in each valve — 42

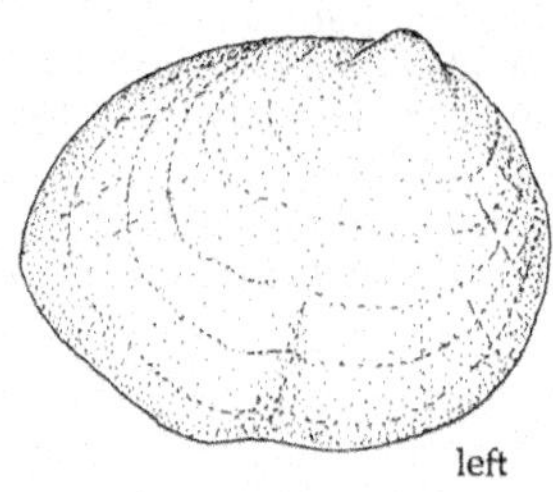

VIII. 44 *Montacuta substriata*

41 A single anterior lateral tooth in each shell valve, no cardinal teeth. Shell with a few fine, radiating ridges, up to 3 mm long; white. At ELWS and sublittoral, in association with the heart urchins *Spatangus purpureus* and *Echinocardium flavescens*. Not common
Montacuta substriata (Montagu) (VIII.44)

– Left valve with one anterior lateral tooth, right valve with one cardinal tooth but no lateral teeth. Shell with fine concentric lines only, up to 8 mm long; dull white with red staining. Lower shore and shallow sublittoral, always in association with the heart urchin *Echinocardium cordatum*
Tellimya ferruginosa (Montagu) (VIII.45)

42 Each valve with one cardinal tooth and single anterior and posterior lateral teeth. Shell with fine radiating striations, less than 2 mm long; yellowish white with a few darker concentric bands. At ELWS and in shallow sublittoral, in association with sipunculids. Not common *Epilepton clarkiae* (Clark) (VIII.46)

– Each valve with single anterior and single posterior lateral teeth; no cardinal teeth. Shell with concentric lines only, up to 3 mm long; white with light brown or olive periostracum. ELWS and sublittoral, often in association with the brittle star *Amphiura brachiata*
Mysella bidentata (Montagu) (VIII.47)

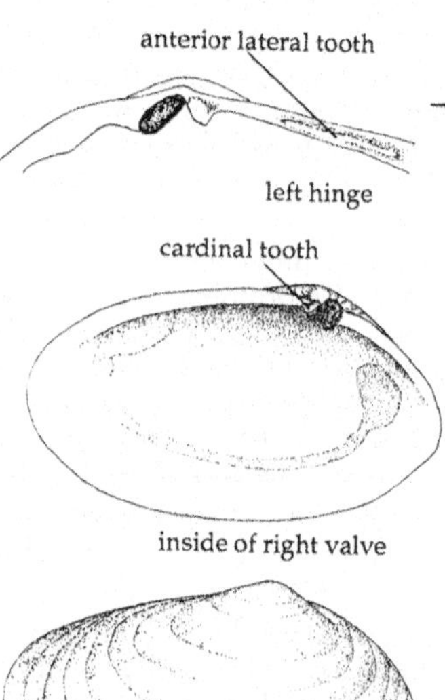

VIII. 45 *Tellimya ferruginosa*

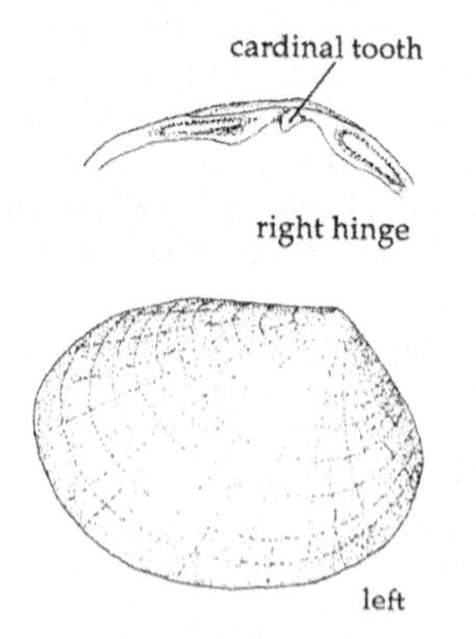

VIII. 46 *Epilepton clarkiae*

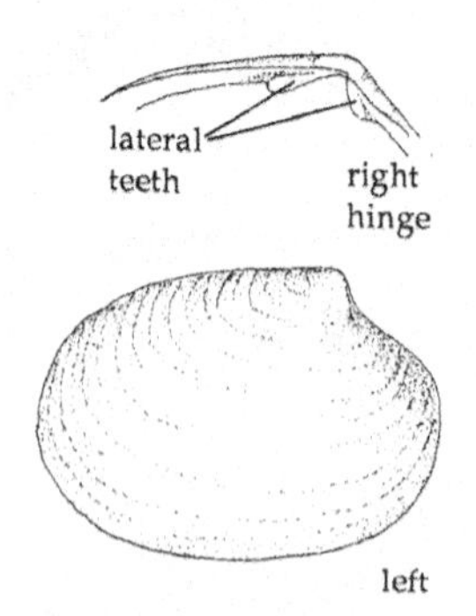

VIII. 47 *Mysella bidentata*

Key IX. Prosobranch gastropods

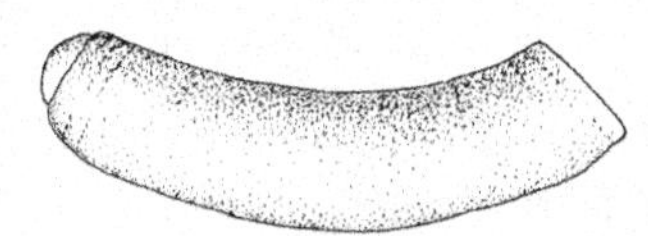

IX. 1 *Caecum glabrum*

All British species of prosobranch snails are keyed and illustrated by Graham (1988). The common periwinkle, *Littorina littorea* (fig. 9), is not an inhabitant of sandy beaches, but is sometimes dislodged from adjacent rocks and stranded.

1 Shell a tiny curved cylinder, less than 5 mm long; one end open, the other sealed with a chalky plug (IX.1) — 2

– Not like this — 3

2 Shell smooth, with a rounded plug. Up to 2.5 mm long. In fine sand or sandy mud. Mostly sublittoral but occasionally at ELWS. South and west coasts only *Caecum glabrum* (Montagu) (IX.1)

– Shell finely ringed, with a pointed plug. Up to 4 mm long. In fine sand or sandy mud. Sublittoral, occasionally at ELWS. South and west coasts only *Caecum imperforatum* Kanmacher (fig.12, p.16)

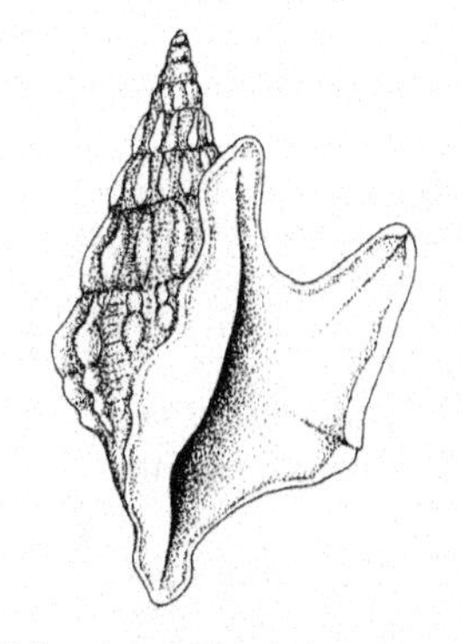

IX. 2 *Aporrhais pespelecani*

3 Shell an elongate spire of about eight ribbed whorls. Aperture (mouth of shell) elongate, with a broad, lobed extension on one side. Sublittoral, on sandy mud, occasionally at ELWS. All British coasts, common pelican's foot shell *Aporrhais pespelecani* (L.) (IX.2)

– Without a broad, lobed extension to the aperture — 4

4 Shell oval, glossy, banded pink and white: the sea slug *Acteon tornatilis* (L.) (pl. 4.2)

– Not pink and white banded — 5

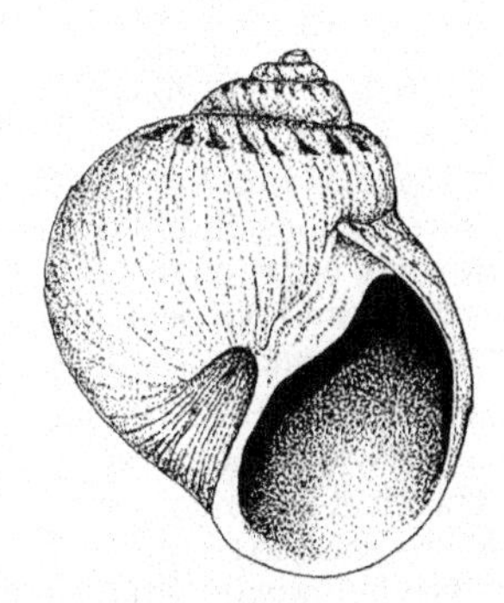

IX. 3 *Lunatia catena*

5 Shell almost globular, with large aperture, the last whorl – body whorl – constituting most of the shell's length (IX.3) — 6

– Shell an elongate spire with body whorl constituting one–third or less of length — 8

6 Shell thin, with squared aperture. Up to 40 mm long, deep violet colour. Pelagic (drifting on the sea surface), occasionally stranded on south and west shores after westerly gales *Janthina janthina* L. (pl. 4.3)

– Shell thick, with oval aperture. Yellowish brown with dark brown markings — 7

IX. 4 *Epitonium clathrus*

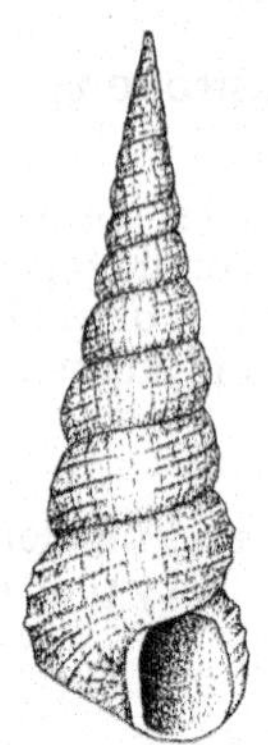
IX. 5 *Turritella communis*

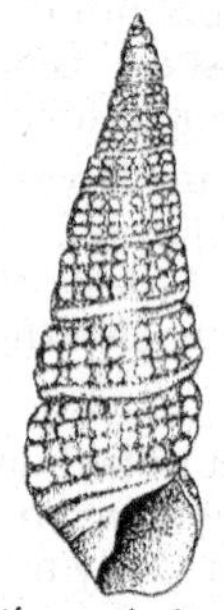
IX. 6 *Bittium reticulatum*

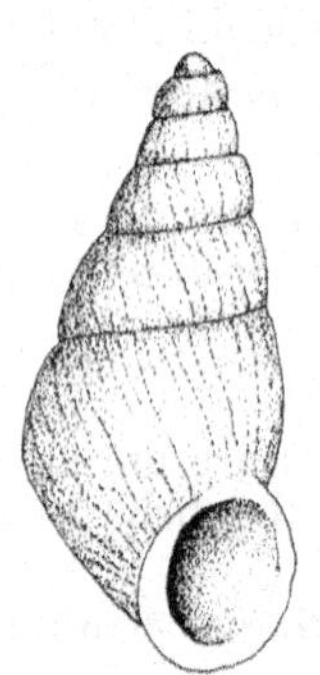
IX. 7 *Hydrobia ulvae*

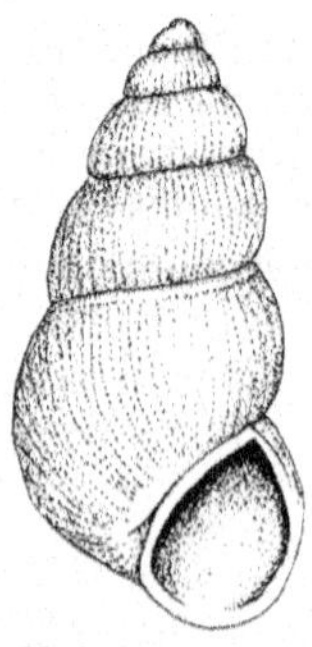
IX. 8 *Hydrobia neglecta*

7 Body whorl with five spiral rows of brown markings. Up to 16 mm long. Lower shore and sublittoral, often seen bulldozing through top layers of sand, leaving a deeply channelled trail. All British coasts, common
Lunatia alderi (Forbes) (pl. 4.4)

– Body whorl with a single spiral row of brown marks, close to its upper edge. Up to 30 mm long. Lower shore and sublittoral; behaviour and feeding as for *L. alderi*. All British coasts, common
Lunatia catena (da Costa) (pl. 4.5;IX.3)

8 Shell smooth, glossy; comprising up to 15 loosely spiralled whorls, not completely touching at edges, crossed by thin, raised ribs. Up to 40 mm long. On sand and sandy mud; sublittoral, but occasionally at ELWS. Off most British coasts but not frequent
Wentletrap *Epitonium clathrus* (L.) (IX.4)

– Shell whorls completely fused along boundaries 9

9 Shell with ribs or spiral ridges. 10 mm or longer 10

– Shell smooth, usually less than 6 mm long, with a low spire of about six whorls 11

10 Shell a slender, sharply pointed cone of up to 20 whorls; with fine spiral ridges but no ribs. Mouth more or less square, its lip without a notch. Up to 55 mm long; light greyish brown to buff, with pale striations. Sublittoral, often abundant on fine muddy sand, occasionally at ELWS. Southern and western coasts, and NE Scotland
Turritella communis Risso (IX.5)

– Shell a blunt spire of up to 16 whorls; with thick ribs and low spiral ridges, forming tubercles where they cross. Aperture oval, with a rounded notch at its inner, lower corner. Up to 15 mm long; light brownish to deeper chestnut, paler on ribs and ridges. On fine silty sand, often in seagrass beds; lower shore and sublittoral. All British coasts, common *Bittium reticulatum* da Costa (IX.6)

11 Shell whorls flat-sided in profile. Head tentacles with black rings near tips. Up to 6 mm long; dull white when dead, yellowish brown when alive. In mud and muddy sand on very sheltered beaches, especially where subject to brackish water. Intertidal, most common above MTL, occasionally lower shore and shallow sublittoral. All British coasts, common *Hydrobia ulvae* (Pennant) (IX.7)

– Whorls distinctly rounded in profile. Head tentacles with triangular black marks near tips, not rings. Up to 4 mm long; dull white or brownish. In mud and muddy sand, especially in estuarine and brackish waters and among sea grasses. Intertidal. Known from the Channel Islands, W. Scotland, W. Ireland and North Sea coasts of England *Hydrobia neglecta* Muus (IX.8)

Key X. Opisthobranchs: Sea slugs

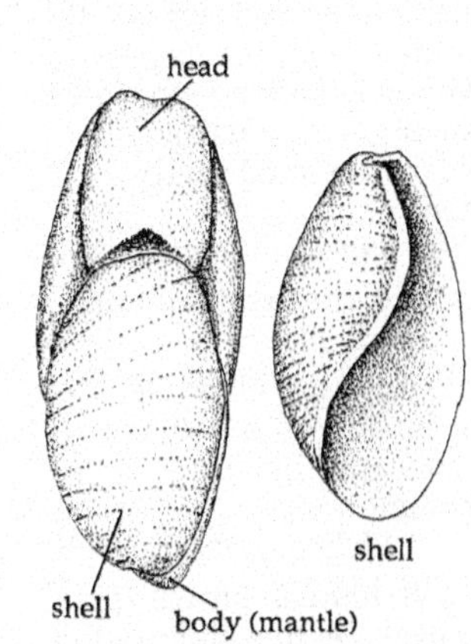

X. 1 *Scaphander lignarius*

Most sea slugs are grazers or predators, and are usually closely associated with their food organisms. A few are infaunal predators, feeding on worms, bivalves and other soft substratum animals, and, unlike most familiar sea slugs, have a shell. All British species are keyed and illustrated by Thompson (1988).

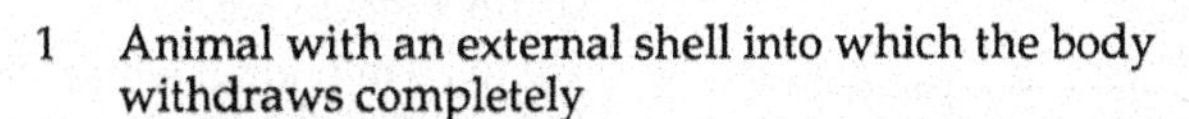

1 Animal with an external shell into which the body withdraws completely 2

– Animal with internal shell only, not visible in active specimens 5

2 Shell thick, with spire of three or four whorls; oval, glossy, banded pink and white; up to 25 mm long, body whorl comprising two-thirds total length. Burrows in fine sand. From MTL into shallow sublittoral, often active at low tide, leaving a channelled trail. South and west coasts, not common *Acteon tornatilis* (L.) (pl. 4.2)

– Shell not banded; with widely flared opening, lacking a spire, or with just two low whorls 3

3 Head end of animal with pair of thick tentacles on rear edge (X.2). Shell thin and fragile, white 4

– Head end of animal lobed at rear but without tentacles. Shell thick, glossy, green yellow or brown; up to 60 mm long. On muddy sand, sublittoral but occasionally at ELWS. South and west coasts only *Scaphander lignarius* (L.) (X.1)

4 Opening of shell shorter than body whorl. Up to 15 mm long (shell 10 mm), white. In mud and muddy sand. From MTL into sublittoral. All British coasts, locally abundant *Retusa obtusa* (Montagu) (X.2)

– Opening of shell longer than body whorl. Up to 7 mm long (shell 5 mm), white or yellowish. In mud and muddy sand, lower shore and shallow sublittoral. All British coasts except southeast *Retusa truncatula* (Bruguière) (X.3)

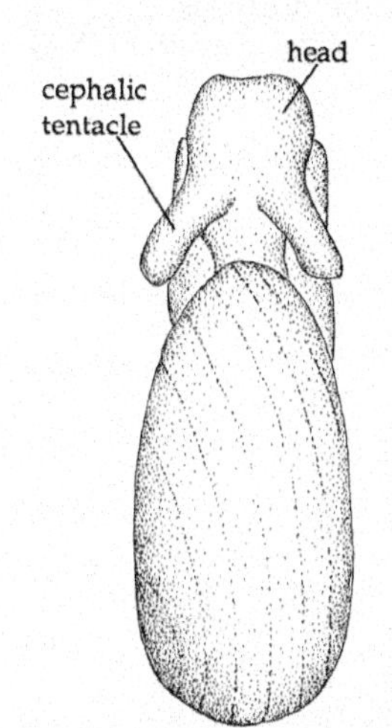

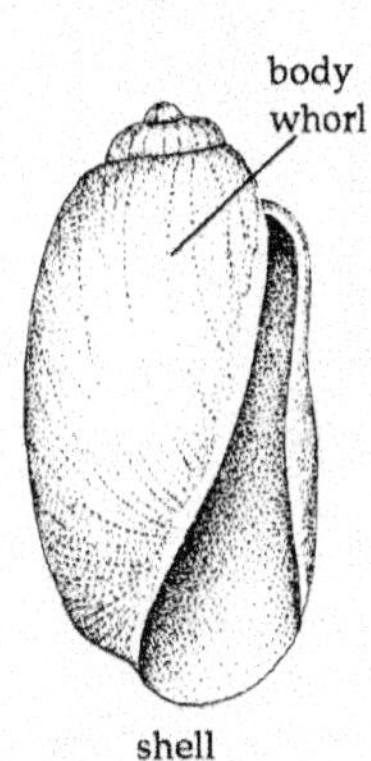

X. 2 *Retusa obtusa*

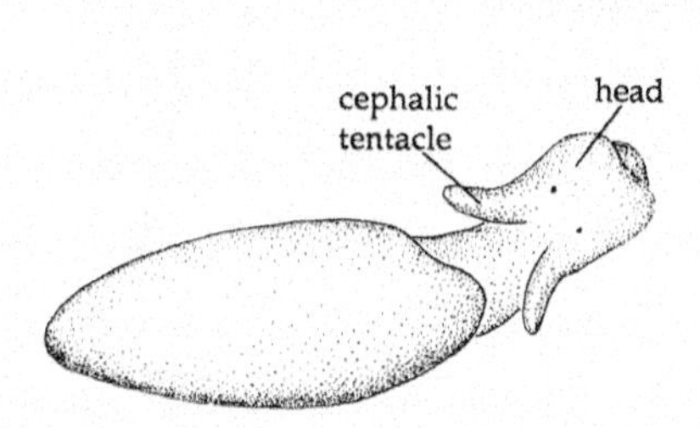

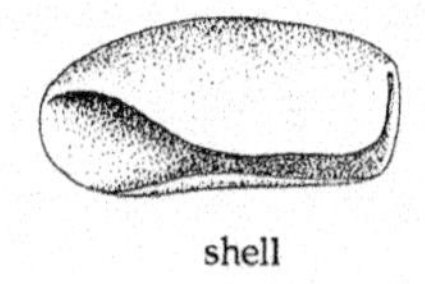

X. 3 *Retusa truncatula*

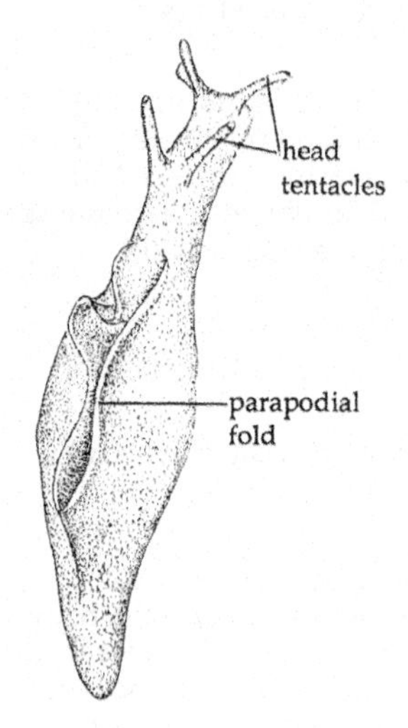

X. 4 *Aplysia punctata*

5 Body of animal with four thick, inrolled tentacles at head end, and a long parapodial fold on each side (X.4). Up to 300 mm long; juveniles pinkish red, adults brown, olive green, purple to black, with blotches and streaks of darker and lighter pigment. Among seaweeds and seagrasses, occasionally at LWS on sheltered sandy mud. All British coasts, common.
Sea Hare *Aplysia punctata* (Cuvier) (X.4)

– Body of animal, viewed from above, consisting of four distinct lobes (X.5) 6

6 Up to 70 mm long; white to pale yellow, with white flecks. Shell shallowly concave, broadly rounded. On fine and muddy sand, lower shore and shallow sublittoral. All British coasts, common
Philine aperta (L.) (X.5)

– Up to 5 mm long, pale yellow with reddish brown spots. Shell deeply concave, elongate oval. On fine and muddy sand, lower shore and shallow sublittoral. All British coasts but most common in north and northeast
Philine punctata (Adams) (X.6)

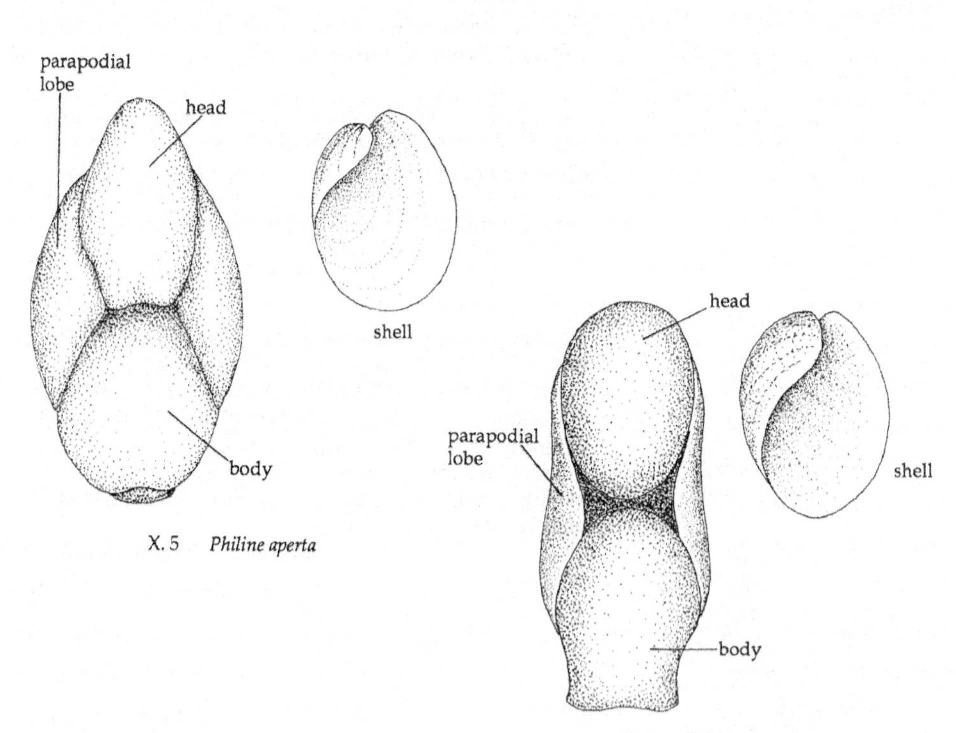

X. 5 *Philine aperta*

X. 6 *Philine punctata*

Key XI. Echinoderms

All British echinoderms are described by Mortensen (1927). A more recent, but not comprehensive, account is given by Moyse and Tyler (1990).

1 **Starfishes**, with broad arms 3–5 times as long as central disc diameter 2
– **Brittle stars**, with slender arms up to 10 times as long as central disc diameter 4
– **Heart urchins** 6
– **Sea cucumbers**, elongate, flaccid, with sticky tube feet 8

2 With five short, triangular arms 3
– With seven long, narrow, floppy arms, often of different lengths, sharply tapered at tips. Diameter up to 600 mm; reddish brown. Sublittoral, occasionally at ELWS. All British coasts, common *Luidia ciliaris* (Philippi) (pl. 8.1)
[*L. sarsi* Duben and Koren has five narrow, parallel-sided arms. It is as common as *L. ciliaris* but more strictly sublittoral, on muddy sand.]

3 Arms stiff, flat; each with a border of small plates. Diameter often exceeding 100 mm; yellowish brown to pale violet. On all but most exposed shores, partly buried in topmost cm of sand; lower shore and sublittoral. All British coasts, common *Astropecten irregularis* (Pennant) (pl. 6.4)
– Arms swollen, not flat; surface warty, without a border of plates. Up to 500 mm diameter; orange, red, purple or yellowish, often mottled. Common on all substrata, often stranded on sandy shores *Asterias rubens* L. (pl. 6.1)

4 Arms very long, thin and curling, with rows of erect spines. Upper surface of disc with small, fine scales, and two large plates close to the base of each arm. Disc up to 12 mm diameter; brownish grey. Lower shore and shallow sublittoral, buried in the sand, with the arm tips protruding. Western coasts only *Amphiura brachiata* (Montagu) (pl. 6.3)
– Arms stouter, rather stiff, with the rows of spines lying flat against surface. Upper surface of disc with large and small scales 5

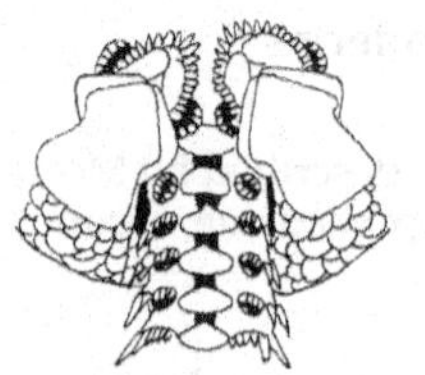

XI. 1 *Ophiura ophiura*, base of arm and part of disc seen from beneath

5 Middle plates on underside of arm, close to its base, separated by pores (XI.1). Upper surface of disc reddish brown, underside paler; up to 35 mm diameter. At LWS and in shallow sublittoral. All British coasts, common *Ophiura ophiura* (L.) (pl. 6.6)

– Middle plates on underside of arm, close to its base, in continuous series, not separated by pores (XI.2). Upper surface reddish brown, underside paler; disc diameter up to 12 mm. At LWS and in shallow sublittoral. All British coasts, common *Ophiura albida* Forbes (XI.2)

XI. 2 *Ophiura albida*, base of arm and part of disc seen from beneath

6 Mouth centrally situated on underside of animal, with teeth visible when living. Upper surface with faint, radiating grooves. Up to 15 mm long, oval and flattened; greenish grey, with a fur of fine spines. In medium to coarse sand and gravel, lower shore and sublittoral. All British coasts, common but inconspicuous *Echinocyamus pusillus* (Müller) (XI.3)

– Mouth situated towards anterior, on underside of animal (XI.4), overhung by a lip, without teeth. Upper surface with deep, petal-shaped grooves 7

7 Purple; with a single ring of long spines around anus. Up to 120 mm long. A shallow burrower – often partly uncovered – in medium to coarse sand, and fine gravel. Mostly sublittoral, often at LWS on southwest coasts. All British coasts, locally common *Spatangus purpureus* Müller (pl. 6.5)

– Light brown, with bright yellow spines; with a double ring of long spines around anus. Up to 90 mm long. A deep burrower – found only by digging – in fine to medium sand. Lower shore and shallow sublittoral. All British coasts, common *Echinocardium cordatum* (Pennant) (XI.4; pl. 6.2)

8 Only one species of sea cucumber is at all frequent on sandy shores: slender, flaccid, pink, up to 300 mm long, with 12 short, retractile tentacles surrounding the mouth. Lower shore and shallow sublittoral, in fine silty sand; adheres readily to bare feet. West coasts from Shetland to Plymouth, east coast as far south as Northumberland *Leptosynapta inhaerens* (Müller) (pl. 7.4)

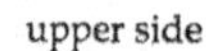

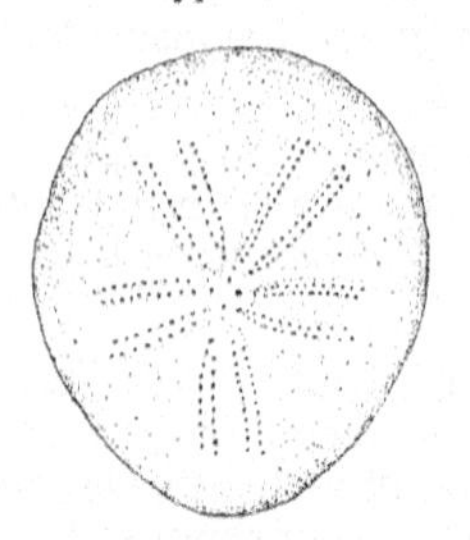

lower side

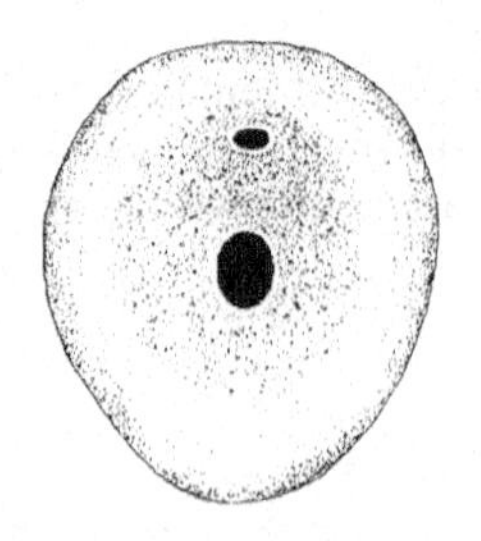

IX. 3 *Echinocyamus pusillus*, test with spines removed

upper side

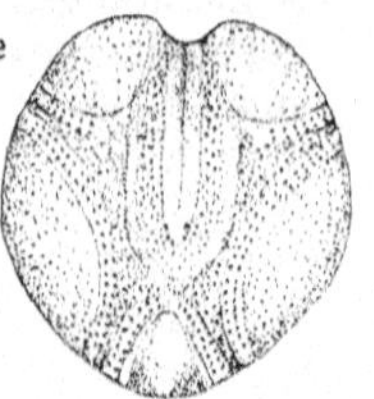

lower side

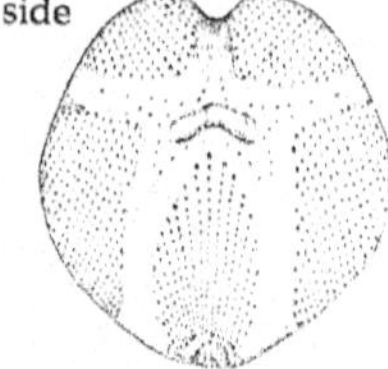

XI. 4 *Echinocardium cordatum*, test with spines removed

7 Techniques

Sandy shore ecology requires relatively unsophisticated field equipment. A broad-tined fork is useful for exploratory digging, but a spade and a box sieve are necessary for accurate sampling. The sieve is simply a stout wooden box about 30x30 cm, and 25 cm deep, with a rigid 1 mm^2 steel mesh bottom. Sand dug from a measured quadrat area should be shovelled directly into the box, which is then sieved by gently shaking it up and down in the water which, below MTL, always collects in the hole. Once all sand has been washed from the sieve its remaining contents should be emptied into a large white plastic tray. The catch may be bagged in its entirety, knotted into a plastic bag with a small quantity of clean seawater and a data label, or selected specimens may be placed in individual tubes, and returned to the laboratory for examination. Shrimps, swimming crabs and small fish are best caught with a shrimp net in ankle-depth water at low tide, and the best results are obtained after dark. Fine mesh nets will catch mysids and the larger amphipods, and a simple conical plankton net, towed across the beach in half a metre of water, will capture even the smallest crustaceans. The Coleman-Segrove surf plankton net (p. 24) is a large and expensive piece of equipment and only likely to be found at a professional marine biological station.

The amphipods, isopods and cumaceans which will form the bulk of each sample are often so abundant across the shore that periodic sampling will have no appreciable effect on their numbers. Each population will comprise a number of overlapping, short-lived generations, and breeding females tend to produce a succession of broods through the reproductive season. However, bivalves and echinoderms are long-lived animals which often do not reproduce successfully every year and it is best not to remove large numbers of these. Most of the larger bivalves, and all of the echinoderms, are readily identified in the field, and accurate data on size, density and age can be collected without removing the animals from the beach. In all cases wasteful sampling should be avoided, and once laboratory examination of material has been completed it is good practice to return living specimens to the beach. It is often helpful to build up a reference collection of named specimens as an aid to identification. Bivalve shells are best kept dry, after washing in fresh water, but all the smaller crustaceans should be preserved in snap-top vials of 70% alcohol. Soft-bodied animals such as polychaetes need to be relaxed in 7% magnesium chloride in seawater, before being initially preserved in 5% seawater formalin and then stored in fresh 5% seawater formalin. Formalin gives off unpleasant and harmful fumes which may be dangerous in poorly ventilated conditions. An alternative preservative is a 1% solution of propylene phenoxetol (made up in hot water) but this can only be used for storage after initial preservation in formalin. Precise data labels should be prepared for each specimen, using Indian ink on goatskin parchment or plastic surfaced paper; these will last indefinitely in all preservative fluids. Other types of paper will suffice for temporary labels, but will eventually disintegrate and thus need to be replaced at regular intervals. If permanent interior labels are not available, it may be best to stick labels to the outsides of containers. Methods of fixation and preservation for different marine invertebrate groups are given in detail by Smaldon and Lee (1979).

Much basic recording work can be achieved on sandy shores simply by collecting and identifying specimens as the tides allow. Similarly, successful ecological studies, even including population dynamics, can be carried out on known populations on particular areas of a beach. However, as a general rule many aspects of sandy shore ecology are best carried out within the framework of a planned transect survey. A transect is simply a straight line, in this case running from LW to HW at a right angle to the water's edge. The top and bottom points are marked with surveying poles, and their positions fixed on a large-scale map by means of a compass. Sampling points are established at measured intervals along the transect line, and the vertical rise between successive stations is recorded using a surveyor's level. Using these data, and the predicted LW level obtained from tide tables, the profile of the beach may be drawn. Admiralty Tide Tables (published by HMSO) are kept at all marine laboratories and field centres, and are often to be found in the main public library of the larger coastal towns. Additionally, locally published tide tables are available in bookshops in many coastal districts. An alternative method of constructing a beach profile is to establish a lowest transect station on the water's edge at the predicted time of low water, and then to record the exact time at which the incoming tide reaches each successive sampling station (remember to measure the distance between each one and the next). The predicted tidal curve for the locality will show what proportion of the total tidal rise will have occurred at successive time intervals, and the vertical rise between successive stations can thus be calculated. Full details of this method, together with worked examples, are given in the Admiralty Tide Tables.

Permanent transects are useful when long-term surveys are being conducted. Each sampling point is simply marked with a house brick, with a 2 m length of thick polypropylene rope attached to it. The brick should be buried 1 m deep; although the level of the sand may vary through the year, enough rope will be showing, provided it is not removed by curious holiday makers, for each sampling point to be easily located. Quantitative data can be collected at each sampling point using standard wire quadrats to define the area to be excavated to a measured depth, or by means of a steel box quadrat, the entire contents of which are removed and sieved once it has been hammered into the sand to a predetermined depth. Samples of sand, for particle size analysis, or small cores for the extraction of meiofauna and microfauna, can be collected using a convenient tube of known diameter and length - plastic domestic plumbing pipe is ideal for these purposes. Whatever sampling procedures are adopted, quantitative methods, involving replicated sampling, are strongly recommended as they provide solid bases for ecological comparisons across, along and between shores, or from year to year, and also the kind of data necessary for population studies.

Sand samples intended for particle size analysis should be dried to constant weight, crushed gently by means of a pestle and mortar to separate the grains, and passed through a graded series of Endecott sieves, the meshes of each of which correspond to the intervals of the Wentworth Scale (table 1). The weight of sand retained by each sieve is expressed as a proportion of the total and a cumulative frequency curve is constructed. Endecott sieves are part of the equipment of all marine stations and field centres.

The identification of most specimens collected on sandy shore transects requires a good binocular microscope. Experimental study of reproductive patterns, feeding or behaviour also often requires sophisticated equipment, and such work is usually only possible at properly equipped field centres. However, it is possible to make original observations on, for example, swimming behaviour, rates of burrowing, prey selection and reproduction on a wide range of animals kept in small glass tanks floored with sand. The Field Studies Council maintains several first-class field centres, such as Dale Fort or Orielton, Pembrokeshire, and Slapton Ley, Devon, which offer courses on marine ecology, and also provide facilities for independent work. Some university-managed centres, such as the Millport Marine Station, Cumbrae, offer occasional courses for non-university classes, while both the Marine Biological Association of the U.K., at Plymouth, and the Scottish Marine Biological Station, Millport, provide bench space for members.

Publication of original work is often possible through local natural history societies, some of which publish periodical journals. Field Studies, the journal of the Field Studies Council, also publishes original regional survey investigations. There are many professional journals, the most important of which are the Journal of the Marine Biological Association of the U.K., the Journal of Experimental Marine Biology and Ecology, and Marine Biology. Manuscripts should only be submitted to these after discussion with a professional biologist.

Increasing environmental awareness has given rise to many local pressure groups, some of which are actively involved in survey or monitoring exercises. These are often managed by county museums, local naturalists' trusts, or nationwide organisations such as the Marine Conservation Society. All of these organizations welcome new members keen to contribute to their programmes, and will readily provide information on the conservation status of their local beaches. Good original fieldwork, including baseline surveys and long-term monitoring projects, can make a valuable contribution to conservation planning at both local and national level. The work of the amateur enthusiast has always been important in British natural history, and it is now even more so if we are to assess the impact of human disturbance, pollution and development on threatened habitats.

Some useful addresses

Marine Biological Association of the United Kingdom,
The Laboratory, Citadel Hill, Plymouth, PL1 2PB.

Scottish Marine Biological Association,
Dunstaffnage Marine Research Laboratory,
P.O. Box 3, Oban, Argyll, PA34 4AD.

The University Marine Biological Station,
Millport, Isle of Cumbrae, KA28 OE9.

Information about Field Studies Council field courses is available from:

The Information Officer, Field Studies Council,
Preston Montford, Montford Bridge,
Shrewsbury, SY4 1HW.

Addresses of local naturalists' trusts are available from:

Royal Society for Nature Conservation, Witham Park,
Waterside South, Lincoln LN5 7JR.

AIDGAP Keys are available from:

The Richmond Publishing Co. Ltd,
P.O. Box 963, Slough, SL2 3RS.

References and further reading

Finding books and journal articles

Some of the books and journals listed below will not be available in school or public libraries. It is sometimes possible to obtain permission to use local university libraries, and main public libraries are usually able to borrow books, or obtain photocopies of scientific papers, from the British Library, Document Supply Centre. The process of finding a reference is made easier if it is correctly cited, as below, with the author's name, followed by the date, title of the article or book, the journal title (or publisher in the case of a book), volume number and page numbers.

References

Admiralty Tide Tables. Vol.1. European Waters. Hydrographer of the Navy, Taunton (HMSO).

Ansell, A.D. (1961). Reproduction, growth and mortality of *Venus striatula* (da Costa) in Kames Bay, Millport. *Journal of the Marine Biological Association of the U.K.*, 41, 191-215.

Ansell, A.D. (1972). Distribution, growth and seasonal changes in biochemical composition for the bivalve *Donax vittatus* (da Costa) from Kames Bay, Millport. *Journal of Experimental Marine Biology and Ecology*, 10, 137-150.

Ansell, A.D. & Bodoy, A. (1979). Comparison of events in the seasonal cycle for *Donax vittatus* and *D. trunculus* in European waters. In: Naylor, E. & Hartnoll, R. (editors), *Cyclic Phenomena in Marine Plants and Animals*. London: Pergamon Press.

Ansell, A.D. & Trevalliou, A. (1967). Studies on *Tellina tenuis* da Costa. 1. Seasonal growth and biochemical cycle. *Journal of Experimental Marine Biology and Ecology*, 1, 220-235.

Barrett, J.H. & Yonge, C.M. (1958). *Collins Pocket Guide to the Sea Shore*. London: Collins (with successive reprints).

Beukema, J.J. (1987). Influence of the predatory polychaete *Nephtys hombergi* on the abundance of other polychaetes. *Marine Ecology Progress Series*, 40, 95-107.

Beukema, J.J. & de Vlas, J. (1979). Population parameters of the Lugworm, *Arenicola marina*, living on tidal flats in the Dutch Wadden Sea. *Netherlands Journal of Sea Research*, 13, 331-353.

Brady, J. (1982). (Editor). *Biological timekeeping*. Cambridge University Press. 197pp.

Brafield, A.E. (1978). *Life in Sandy Shores*. London: Edward Arnold Ltd.

Brown, A.C. & McLachlan, A. (1990). *Ecology of Sandy Shores*. Amsterdam: Elsevier Science Publishers B.V. 328pp.

Cadman, P.S. & Nelson-Smith, A. (1993). A new species of Lugworm: *Arenicola defodiens* sp. nov. *Journal of the Marine Biological Association of the U.K.*, 73, 213-223.

Clausen, C. (1967). Morphological studies of *Halammohydra* Remane (Hydrozoa). *Sarsia*, 29, 349-370.

Coleman, J.S. & Segrove, F. (1955). The tidal plankton over Stoupe Beck Sands, Robin Hood's Bay (Yorkshire, North Riding). *Journal of Animal Ecology*, 24, 445-462.

Crothers, J. & Crothers, M. (1983). A Key to the Crabs and crab-like animals of British inshore waters. *Field Studies*, 5, 753-806.

Dale, N.G. (1974). Bacteria in intertidal sediments: factors related to their distribution. *Limnology and Oceanography*, 19, 509-518.

Dales, R.P. (1958). Survival of anaerobic periods by two intertidal polychaetes, *Arenicola marina* (L.) and *Owenia fusiformis* Delle Chiaje. *Journal of the Marine Biological Association of the U.K.*, 37, 521-529.

De Vlas, J. (1979). Secondary production by tail regeneration in a tidal flat population of lugworms (*Arenicola marina*), cropped by flatfish. *Netherlands Journal of Sea Research*, 13, 362-393.

De Wilde, P.A.W.J. & Berghuis, E.M. (1979). Spawning and gamete production in *Arenicola marina* in the Netherlands Wadden Sea. *Netherlands Journal of Sea Research*, 13, 503-511.

Edwards, J.M. & Naylor, E. (1987). Endogenous circadian changes in orientational behaviour of *Talitrus saltator*. *Journal of the Marine Biological Asociation of the U.K.*, 67, 17-26.

Edwards, R.R.C., Steele, J.H. & Trevalliou, A. (1970). The ecology of O-group plaice and common dabs in Loch Ewe. III. Prey-predator experiments with plaice. *Journal of Experimental Marine Biology and Ecology*, 4, 156-173.
Elmhirst, R. (1931). Studies on the Scottish marine fauna. The Crustacea of the sandy and muddy areas of the tidal zone. *Proceedings of the Royal Society of Edinburgh*, 51, 169-175.
Eltringham, S.K. (1971). *Life in mud and sand*. London: English Universities Press. 218pp.
Farke, H. & Berghuis, E.M. (1979). Spawning, larval development and migration behaviour of *Arenicola marina* in the laboratory. *Netherlands Journal of Sea Research*, 13, 512-528.
Fincham, A.A. (1970). Amphipods in the surf plankton. *Journal of the Marine Biological Association of the U.K.*, 50, 177-198.
Fish, J.D. & Fish, S. (1972). The swimming rhythm of *Eurydice pulchra* Leach and a possible explanation of intertidal migration. *Journal of Experimental Marine Biology and Ecology*, 8, 195-200.
Fish, J.D. & Preece, G.S. (1970). The annual reproductive patterns of *Bathyporeia pilosa* and *B. pelagica*. *Journal of the Marine Biological Association of the U.K.*, 50, 475-488.
Fish, S. (1970). The biology of *Eurydice pulchra*. *Journal of the Marine Biological Association of the U.K.*, 50, 753-768.
George, J.D. & Hartmann-Schröder, G. (1985). Polychaetes: British Amphinomida, Spintherida and Eunicida. *Linnean Society Synopses of the British Fauna* (n.s.), 32, 1-221.
Gibbs, P.E. (1977). British Sipunculans. Linnean Society Synopses of the British Fauna (n.s.), 12, 1-35.
Gibson, R. (1982). British Nemerteans. *Linnean Society Synopses of the British Fauna* (n.s.), 24, 1-212.
Graham, A. (1988) Molluscs: prosobranch and pyramidellid gastropods. *Linnean Society Synopses of the British Fauna* (n.s.), 2 (2nd ed.), 1-662.
Guillou, J. & Sauriau, P.G. (1985). Some observations on the biology and the ecology of a *Venus striatula* population in the bay of Douarnenez, Brittany. *Journal of the Marine Biological Association of the U.K.*, 65, 889-900.
Harris, R.P. (1972). Horizontal and vertical distribution of the interstitial harpacticoid copepods of a sandy beach. *Journal of the Marine Biological Association of the U.K.*, 52, 375-387.
Hayward, P.J. (1988). *Animals on Seaweed*. Naturalists' Handbooks, no.9. Slough, The Richmond Publishing Co. Ltd, 108pp.
Hayward, P.J. & Ryland, J.S. (1990). *The Marine Fauna of the British Isles and North-west Europe*. Oxford, Clarendon Press, 2 volumes.
Hicks, G.R. & Coull, B.C. (1983). The ecology of marine meiobenthic harpacticoid copepods. *Oceanography and Marine Biology Annual Review*, 21, 67-175.
Higgins, R.P. & Thiel, H. (1988). *Introduction to the study of meiofauna*. Washington: Smithsonian Institution Press. 488pp.
Holme, N.A. & McIntyre, A.D. (editors). (1984). *Methods for the study of marine benthos*. 2nd Edition. Blackwell Scientific Publications, Oxford. 387pp.
Ingle, R.W. (1980). *British Crabs*. Oxford and London: Oxford University Press and British Museum (Natural History).
Ingle, R.W. (1983). Shallow water crabs. *Linnean Society Synopses of the British Fauna* (n.s.), 25, 1-206.
Isaac, M.J. (1990). Copepoda. In: Hayward, P.J. and Ryland, J.S. (editors), *The Marine Fauna of the British Isles and North-West Europe*. Vol.1. Ch.8. Oxford, Clarendon Press.
Jensen, K.T. (1985). The presence of the bivalve *Cerastoderma edule* affects migration, survival and reproduction of the amphipod *Corophium volutator*. *Marine Ecology Progress Series*, 25, 269-277.
Jones, A.M. (1979). Structure and growth of a high level population of *Cerastoderma edule* (Lamellibranchiata). *Journal of the Marine Biological Association of the U.K.*, 59, 277-287.
Jones, D.A. (1970). Population densities and breeding in *Eurydice pulchra* and *Eurydice affinis* in Britain. *Journal of the Marine Biological Association of the U.K.*, 50, 635-655.
Jones, D.A. & Naylor, E. (1970). The swimming rhythm of the sand beach isopod *Eurydice pulchra*. *Journal of Experimental Marine Biology and Ecology*, 4, 188-199.
Jones, N.S. (1976). British Cumaceans. *Linnean Society Synopses of the British Fauna*, (n.s.), 7, 1-66.
Jones, W.E. (1980). Field teaching methods in shore ecology. In: Price, J.H., Irvine, D.E.G. & Farnham, W.F. *The Shore Environment, Vol.1: Methods*. London: Academic Press, for the Systematics Association.
Le Mao, P. (1986). Feeding relationships between the benthic infauna and the dominant benthic fish of the Rance Estuary (France). *Journal of the Marine Biological Association of the U.K.*, 66, 391-401.

Lincoln, R.J. (1979). *British Gammaridean Amphipods.* London, British Museum (Natural History). 658pp.
Longbottom, M.R. (1970). The distribution of *Arenicola marina* (L.) with particular reference to the effects of particle size and organic matter of the sediments. *Journal of Experimental Marine Biology and Ecology*, 5, 138-157.
Makings, P. (1977). A guide to the British coastal Mysidacea. *Field Studies*, 4, 575-595.
Manuel, R.L. (1983). *The Anthozoa of the British Isles - a colour guide.* 2nd Edition. Produced for the Marine Conservation Society by R. Earll. Marine Conservation Society, Ross-on-Wye.
Manuel, R.L. (1988). British Anthozoa. *Linnean Society Synopses of the British Fauna* (n.s.), 18, 1-241. Revised edition.
McIntyre, A.D. (1969). Ecology of marine meiobenthos. *Biological Reviews*, 44, 245-290.
Moore, C.G. (1979*a*). The distribution and ecology of psammolittoral meiofauna around the Isle of Man. *Cahiers de Biologie marine*, 20, 383-415.
Moore, C.G. (1979*b*). The zonation of psammolittoral harpacticoid copepods around the Isle of Man. *Journal of the Marine Biological Association of the U.K.*, 59, 711-724.
Morgan, C.E. & King, P.E. (1976). British Tardigrades. *Linnean Society Synopses of the British Fauna* (n.s.), 9, 1-133.
Morgan, E. (1973). On the pressure response of *Eurydice*. *Marine Behaviour and Physiology*, 1, 323-339.
Mortensen, T. (1927). *Handbook of the Echinoderms of the British Isles.* Oxford University Press.
Moyse, J. & Tyler, P.A. (1990). Echinodermata. In: Hayward, P.J. and Ryland, J.S. (editors). *The Marine Fauna of the British Isles and North-west Europe*, Vol.2, ch.15. Oxford, Clarendon Press.
Naylor, E. (1972). British Marine Isopods. *Linnean Society Synopses of the British Fauna* (n.s.), 3, 1-86.
Naylor, E. (1990). Isopoda. In: Hayward, P.J. & Ryland, J.S. (editors). *The Marine Fauna of the British Isles and North-west Europe*, Vol.1, Ch.9. Oxford, Clarendon Press.
Nelson-Smith, A & Knight-Jones, P. (1990). Annelida: Polychaeta. In: Hayward, P.J. and Ryland, J.S. (editors). *The Marine Fauna of the British Isles and North-west Europe*, Vol.1, Ch.6. Oxford, Clarendon Press.
Newell, G.E. (1948). A contribution to our knowledge of the life history of *Arenicola marina* L. *Journal of the Marine Biological Association of the U.K.*, **28**, 554-580.
Pleijel, F. & Dales, R.P. (1991). Polychaetes: British Phyllodocoideans, Typhloscolecoideans and Tomopteroideans. *Synopses of the British Fauna* (n.s.), 45, 1-202.
Preece, G.S. (1971). The swimming rhythm of *Bathyporeia pilosa* (Crustacea: Amphipoda). *Journal of the Marine Biological Association of the U.K.*, 51, 777-791.
Reid, D.G. & Naylor, E. (1985). Free-running, endogenous semilunar rhythmicity in a marine isopod crustacean. *Journal of the Marine Biological Association of the U.K.*, 65, 85-91.
Ruppert, E.E. (1988). Gastrotricha. pp. 302-311 in: Higgins, R.P. & Thiel, H. *Introduction to the study of Meiofauna*. Smithsonian Institution, Washington and London.
Sars, G.O. (1899). *An account of the Crustacea of Norway.* Vol.2. *Isopoda.* Bergen Museum, Bergen, 270pp.
Sars, G.O. (1911). *An account of the Crustacea of Norway.* Vol.5. *Copepoda, Harpacticoida.* Bergen Museum, Bergen. 449pp.
Smaldon, G. (1979). British coastal shrimps and prawns. *Linnean Society Synopses of the British Fauna* (n.s.), 15, 1-126.
Smaldon, G. & Lee, E.W. (1979). A synopsis of methods for the narcotization of marine invertebrates. *Royal Scottish Museum Information Series, Natural History*, 6, 96pp.
Steers, J.A. (1962). *The Sea Coast.* 3rd Edition. New Naturalist Series, London: Collins.
Steers, J.A. (1969*a*). *Coasts and Beaches.* Contemporary Science Paperbacks, 34, Edinburgh: Oliver and Boyd.
Steers, J.A. (1969*b*). *The Coastline of England and Wales.* 3rd Edition, reprinted. Cambridge University Press.
Swedmark, B. (1964). The interstitial fauna of marine sand. *Biological Reviews*, 39, 1-42.
Tattersall, W.M. & Tattersall, O.S. (1951). *The British Mysidacea.* London, The Ray Society. 460pp.
Tebble, N. (1976). *British bivalve seashells.* 2nd Edition. Edinburgh, Her Majesty's Stationery Office.
Thompson, T.E. (1988). Molluscs: Benthic Opisthobranchs. *Linnean Society Synopses of the British Fauna* (n.s.), 8, 1-356. 2nd Edition.
Trueman, E.R., Brand, A.R. and Davis, P. (1966). The dynamics of burrowing of some littoral bivalves. *Journal of Experimental Biology*, 44, 469-492.

Warwick, R.M. (1989). The role of meiofauna in the marine ecosystem: evolutionary considerations. *Zoological Journal of the Linnean Society*, 96, 229-241.
Watkin, E.E. (1939). The pelagic phase in the life history of the amphipod genus *Bathyporeia*. *Journal of the Marine Biological Association of the U.K*, 23, 467-481.
Watkin, E.E. (1941). Observations on the night tidal migrant crustacea of Kames Bay. *Journal of the Marine Biological Association of the U.K.*, 25, 81-96.
Westheide, W. (1990). Polychaetes: Interstitial families. *Synopses of the British Fauna* (n.s.), 44, 1-152.
Williams, J.A. (1978). The annual pattern of reproduction of *Talitrus saltator* (Crustacea: Amphipoda: Talitridae). *Journal of Zoology, London*, 184, 231-244.
Williams, J.A. (1979). A semi-lunar rhythm of locomotor activity and moult synchrony in the sand-beach amphipod *Talitrus saltator*. In Naylor, E. and Hartnoll, R. (editors). *Cyclic phenomena in marine plants and animals*. Pergamon Press, London.
Williams, J.A. (1980). The light response rhythm and seasonal entrainment of the endogenous circadian locomotor of *Talitrus saltator* (Crustacea: Amphipoda). *Journal of the Marine Biological Association of the U.K.*, 60, 773-785.
Williams, J.A. (1983). The endogenous locomotor activity rhythm of four supralittoral peracarid crustaceans. *Journal of the Marine Biological Association of the U.K.*, 63, 481-492.
Wright, J.M. (1983). Sand-dwelling ciliates of South Wales. *Cahiers de Biologie marine*, 24, 187-214.
Yonow, N. (1989). Feeding observations on *Acteon tornatilis* (L.) (Opisthobranchia: Acteonidae). *Journal of Molluscan Studies*, 55, 97-102.

Index